A Rare Breeds Album
of American Livestock

Carolyn J. Christman

D. Phillip Sponenberg

Donald E. Bixby

Foreword by Wendell Berry

The American Livestock Breeds Conservancy

Pittsboro, North Carolina

The American Livestock Breeds Conservancy
PO Box 477, Pittsboro, NC 27312 USA

A Rare Breeds Album of American Livestock

Printed in the United States by Glover Printing, Inc., Raleigh, NC

ISBN 1–887316–02–7

COVER PHOTOS: Gotland Horse - Karen Kachur; Navajo-Churro sheep - Maryanne Mott; San Clemente goat - Richard Kirschman; American Mammoth Jackstock - Mary Ellen Nicholas; Hereford pig - Mark Hess; Dexter cow - Catherine Palladino

Acknowledgements

The writing and production of this book have been supported by the C. S. Fund, the Charles DeVlieg Foundation, the Geraldine R. Dodge Foundation, the W. K. Kellogg Foundation, the Kohlberg Foundation, the Nan Yarnall Trust Fund, and the members of the American Livestock Breeds Conservancy. This volume is the second of ALBC's guides to rare livestock breeds. It follows the general format and goals of the *American Minor Breeds Notebook*, written by Laurie Heise and Carolyn Christman and published in 1989.

Many people have contributed to our knowledge of domestic species and breeds, but a few individuals merit special acknowledgement. Stefan Adalsteinsson has described the concept of "genetic heritage," essential to our arguments for conservation. Juliet Clutton-Brock and Stephen J. G. Hall have written extensively about the historical development of British livestock breeds as well as the reasons for breed conservation today. Gus Cothran has illuminated the mysteries of equine breed character and relationships. Cary Fowler, Pat Mooney, and Hope Shand have described the evolving context for agricultural diversity. Bruce Kalk has documented the process of breed extinction in the United States. Ian Mason has characterized the evolution of domesticated animals and the development of breeds, as well as the origins and relationships of modern breeds. Lawrence Alderson, Elizabeth Cambra, Hans-Peter Grunenfelder, Elizabeth Henson, Arthur Mariante, and Chris Smith are among those who have provided information about the status of international breeds.

The associations that promote rare breeds in the United States and Canada are valued colleagues of the American Livestock Breeds Conservancy. Many breed associations provided information and photographs for this book, and we are grateful for their partnership.

This book would not have been possible without the illustrations contributed by many talented photographers. Deborah Wechsler skillfully wove words and images together for the book design, and Cynthia Ehrman polished it with her editorial skills. Jonathan Greene gave the book its name and provided much welcome advice during its production.

We dedicate this book to the members of the American Livestock Breeds Conservancy.

Foreword

British Sheep, a sort of album published by the National Sheep Association of Britain, contains photographs and descriptions of 65 breeds and "recognized half breeds." I think that this diversity of breeds is one of the greatest accomplishments of agriculture. It shows that in Britain, which is not much bigger than Kansas, thousands of farmers were paying the most discriminating attention, not only to their sheep, but also to their local landscapes and economies, for a long time. As breeders, they were responding intelligently to local conditions and local needs.

Such local adaptation is the most important requirement for agriculture, wherever it occurs. If we are going to fit our farming to our land, as J. Russell Smith instructed us to do, we must recognize the great diversity of agricultural landscapes and climates. To respond intelligently to the diversity of farms, we need a diversity of livestock breeds and of types within breeds. This is one of the justifications of the work of the American Livestock Breeds Conservancy.

The diversity of livestock breeds and domestic plant varieties can be thought of as a sort of vocabulary with which we may make appropriate responses to the diversity of localities. Local adaptation always requires reasonably correct answers to two questions: What is the nature of the place? and, What are the needs and opportunities of the local economy?

Intelligent livestock breeders may find, in practice, that the two questions become one: How can I produce the best meat at the lowest economic and ecological cost? This question cannot be satisfactorily answered by the market or the meat-packing industry, by breed societies or show ring judges, by "animal science" experts or genetic engineers. It can be satisfactorily answered only by the farmer, and only if the farm, the place itself, is allowed to play a part in the process of selection.

It goes without saying that the animal finally produced by any farm will be a product to some extent of the judgement of the farmer, the meat packer, the breed society, and the show ring judge. But the farm too must be permitted to make and enforce its judgement. If it is not permitted to do so, then there can be no local adaptation. Where there is no local adaptation, the farmer and also the farm must pay significant penalties.

In our time, because of commercial demand and the allure of the show ring, livestock breeders have tended to concentrate on the production of outstanding individual animals as determined by the standards of breed societies or packing houses. Little attention has been paid to the issue of local adaptation – which is really to say that little attention has been paid to the cost of production. This negligence is possible only because, in our time, "cheap" fossil fuel has set the pattern in agriculture. Thoughtlessness in breeding has been subsidized by large checks paid to veterinarians and drug companies, and covered over by the fat made of allegedly cheap corn.

Allegedly cheap fossil fuel and feed grains have pushed agriculture toward uniformity, obscuring regional differences and therefore the usefulness of locally adapted breeds, especially those that do well on forages. It follows that the livestock economy is now dominated by only a few breeds, and that those breeds are large and grain-dependent. Nearly all dairy cows, for example, now are Holsteins, and the modern sheep is more than likely to have a black face and to be "big and tall." This is another reason why we need a Livestock Breeds Conservancy.

My friend Maury Telleen has pointed out to me that 50 years ago the Ayrshire was a popular dairy cow in New England and Kansas because of her ability to make milk on the feed that was locally available; she did not require the optimal conditions of Illinois or Iowa. She was, Maury said, "a cow that could get along." That, I think, is the very point of breed conservation. It is dangerous to think that we have got beyond the need for farm animals that can "get along."

If we assume that the inescapable goal of the livestock farmer, especially in the present economy, must be to reduce costs, and, further, that costs are most readily reduced by adapting our animals to local conditions, then we will have recovered one of the essential standards of livestock breeding. Corn, whatever the market price, is not cheap. What is cheap is grass – *grazed* grass – and where the grass grows determines the kind of animal needed to graze it.

Especially now, when grain feeding and confinement feeding are so common, no American breeder should expect any breed to be locally adapted. Breeders should recognize that from the standpoint of local adaptation and cheap production, every purchase of a breeding animal is a gamble. A newly purchased breeding animal may improve the performance of your flock or herd on your farm, or it may not. Good breeders will know, or they will soon find out, that theirs is not the only judgement involved. While the breeder is judging, the breeder's farm is judging as well, enforcing its demands and making selections.

Because of the ultimately inescapable demands of local adaptation – meat and milk from grass, at the lowest cost – livestock production will never be quite the same pursuit as livestock breed conservation. But the need for economically and ecologically sound livestock production certainly does clarify the necessity for breed conservation. Every trait and quality of the most out-of-favor breed, having already proved useful somewhere, must be presumed to be potentially useful somewhere. The industrialists of agriculture have always taken for granted that whatever is not presently useful or profitable may safely be discarded, trusting that human cleverness could always make up the loss if necessary. But that assumption does not rest on proof, and it implies many dangers to the culture of farming and to our food supply.

—*Wendell Berry*

Table of Contents

Acknowledgements iii
Foreword v
Introduction 1
Chapter 1. Why rare breeds? 5
Chapter 2. Asses 7
- American Mammoth Jackstock 8
- Poitou 10

Chapter 3. Cattle 13
- Ankole-Watusi 15
- Ayrshire 17
- Belted Galloway 17
- Canadienne 19
- Devon 20
- Dexter 21
- Dutch Belted 22
- Florida Cracker 24
- Galloway 25
- Guernsey 27
- Highland 28
- Kerry 30
- Milking Devon 31
- Milking Shorthorn 32
- Pineywoods 33
- Randall Lineback 35
- Red Poll 36
- White Park 37

Chapter 4. Goats 39
- Nigerian Dwarf 40
- Oberhasli 42
- San Clemente 43
- Spanish 44
- Tennessee Fainting 45

Chapter 5. Horses 47
- Akhal-Teke 48
- American Cream 49
- Canadian 51
- Caspian 52
- Cleveland Bay 53
- Clydesdale 54
- Dartmoor 56
- Exmoor 57
- Florida Cracker 58

Friesian 59
Gotland 61
Hackney Horse 61
Lipizzan 63
Mountain Pleasure 64
Rocky Mountain 65
Shire 67
Spanish Mustang/Spanish Barb 67
Suffolk 69
Chapter 6. Pigs 71
Choctaw 73
Gloucestershire Old Spots 75
Guinea Hog 75
Hereford 77
Large Black 78
Mulefoot 79
Ossabaw Island 80
Red Wattle 81
Saddleback 82
Tamworth 83
Chapter 7. Sheep 85
Barbados Blackbelly 87
California Variegated Mutant/Romeldale 88
Cotswold 89
Dorset Horn 90
Gulf Coast Native 91
Hog Island 93
Jacob (American) 94
Karakul (American) 95
Katahdin 96
Leicester Longwool 97
Navajo-Churro 98
Oxford 100
Santa Cruz 101
Shetland 102
St. Croix 103
Tunis 104
Wiltshire Horn 105
Chapter 8. Conservation Breeding 107
Selected Bibliography 109
Breed and Species Organizations 110
Photo Credits 115
Index 117

Introduction

THROUGHOUT THE LONG HISTORY of agriculture, countless individuals have dedicated themselves to caring for and breeding domestic animals. In every corner of the world, people selected and bred animals to fit their present needs. Distinct breeds evolved, each reflecting the interplay among people, animals, and environments. Breed diversity within each species today is a legacy from our ancestors and reflects the ways that they have lived around the world.

The range of breeds within each species also represents the genetic diversity that exists in that species. Diversity allows a species to adapt to changing environmental conditions and to respond to a variety of human needs. Consider, for example, three breeds of sheep: the charming Shetland, with its fine, soft wool and adaptation to wet, cold regions; the sleek St. Croix, a hair sheep whose lack of wool allows it to prosper in the tropics; and the elegant Karakul, which thrives in rugged, desert regions and produces durable wool. These are among the hundreds of useful breeds of sheep around the world. Unfortunately, these three breeds and many others are now endangered. The loss of breed diversity endangers the biological health of the sheep species and the human endeavors that make up agriculture. This potential disaster goes far beyond sheep, threatening all of the livestock and poultry species.

The American Livestock Breeds Conservancy (ALBC) is a nonprofit membership organization protecting genetic diversity in livestock and poultry through the conservation of rare breeds in the United States. ALBC produced this book to show what is at stake – the breeds that are in danger of being lost. Some of the breeds included are so rare that they are not often seen in books or periodicals, while others are well known but have become rare without notice. Conservation of rare breeds is necessary if these animals are to remain a part of our world and be a part of our legacy to future generations.

How to Use This Book

A Rare Breeds Album of American Livestock is a guide to endangered breeds of asses, cattle, goats, horses, pigs, and sheep. (The poultry breeds will be described in a separate publication by ALBC.)

The book explains the breed concept and describes why breeds become rare. The chapters on each species open with a general discussion of the uses and status of the species. They then describe the characteristics, history, uses, and prospects for each rare breed. Breed origins, distribution, classification, and status are listed above the text for each breed. The following terms are used.

Origin is the country or region where the breed was developed.

Classification is the type of breed. See the box entitled "Breed Types in North America" for more information.

Distribution lists the countries or regions with major populations of the breed. Breeds found in many countries are considered international. Several breeds are unique to the United States, and any regional concentrations of these are given in parentheses.

Status is described by placing the breed in one of three categories: Critical, Rare, or Watch. The key indicator of status is reproductive performance, and this is measured by the number of annual purebred registrations in North America. The number of registrations approximates the number of purebred animals born that will be kept for breeding. This figure is more accurate than total population estimates, and registration numbers can be compared from year to year to understand trends within a breed. Global population ceilings are included in each category to eliminate breeds which are rare in North America but numerous elsewhere, since such breeds are not threatened with extinction. See the "Conservation Priority Breed List" on page four.

The United States and Canada are considered a single geographical unit since most breed populations are shared between the two countries. Livestock can generally be moved across the border unimpeded, so the exchange of breeding stock has been widely practiced. Most animals are registered only in Canada or in the United States, so it is possible to gather figures for each country and then add these for a continental total.

Critical – Fewer than 200 annual registrations in North America and a global population estimated below 2,000.

Rare – Fewer than 1,000 annual registrations in North America and a global population estimated below 5,000.

Watch – Fewer than 2,500 annual registrations in North America and a global population estimated below 10,000. Also included in this category are breeds which are more popular but which present numerical, geographic, or genetic concerns.

You will encounter what seems like repetition as you read about rare breeds. Many of them are described as having traits related to survival: high fertility, foraging ability, intelligence, longevity, soundness, and adaptation to specific climates or environments. This seeming redundancy is not an accident. Survival traits are relatively unimportant in modern, industrial agriculture. As a result, most of the breeds with these traits have become rare.

The book closes with a discussion of conservation breeding and how to become involved with rare breeds. The list of breed and species organizations provides useful contacts. The Bibliography lists the primary sources for this book.

It is our goal that this book will promote greater understanding and appreciation of rare breeds and encourage their conservation. ❖

Breed Types in North America

Four breed types are found in North America: landraces, standardized breeds, industrial stocks, and feral breeds. Each of these differs according to its development, its selection history, the level of uniformity of the animals, and the role of human breeders.

Landraces are populations which have been shaped by both natural selection and human selection. They are well adapted to specific environments and have excellent reproductive and survival characteristics. Landraces generally have not been selected for production characteristics, and they are more variable in appearance than are standardized breeds, especially in cosmetic points such as color and size. Consistency is the rule, however, in their ability to thrive in demanding environments without human intervention.

Standardized breeds are what most people recognize as breeds: a uniform and predictable group of animals, defined and selected toward a breed standard. Standardized breeds developed from human efforts to organize landraces. They have uniformity of appearance and well developed, consistent production traits. The majority of breeds (and rare breeds) in the United States are standardized.

Industrial stocks originated from a portion of a standardized breed or a cross between two or more standardized breeds which was further selected for maximum performance in a closely controlled environment. "Industrial" refers to the highly organized production systems under which the animals are bred and raised, and the term "stocks" is used because industrial animals may be breeds, crossbreeds, or strains. Industrial stocks are more productive than the standardized breeds which were their parents, but they are less adaptable to other environments. Breeding stock is generally held by a few multinational corporations. Though some strains within these populations may be endangered, information about them is a closely guarded secret, and the animals themselves are generally unavailable to individuals. For these reasons, no industrial stocks are considered rare breeds.

Feral populations result when animals escape or are released from human management and once again live under natural selection. Most feral populations have been extensively crossbred and are therefore not true breeds in the genetic sense. A few feral populations in North America, however, have been isolated for a long time and have come to fit the genetic definition of a breed. Some of these feral breeds are being conserved in their natural habitat (*in situ*) while others are conserved *ex situ* (literally, out of their habitat) as redomesticated animals in the hands of breeders.

ALBC Conservation Priority Breed List

Critical—Fewer than 200 annual North American registrations and estimated fewer than 2,000 global population.

Rare—Fewer than 1,000 annual North American registrations and estimated fewer than 5,000 global population.

Watch—Fewer than 2,500 annual North American registrations and estimated fewer than 10,000 global population. Also included are breeds which present numerical, geographic, or genetic concerns.

	Critical	Rare	Watch
Asses	Poitou	American Mammoth Jackstock*	
Cattle	Canadienne* Devon* Dutch Belted Florida Cracker* Kerry Milking Devon* Pineywoods* Randall Lineback* White Park	Ankole-Watusi Dexter Red Poll	Ayrshire Belted Galloway Galloway Guernsey Highland Milking Shorthorn
Goats	San Clemente (F)*	Nigerian Dwarf * Tennessee Fainting*	Oberhasli Spanish*
Horses	Akhal-Teke American Cream* Canadian* Caspian Cleveland Bay Exmoor Florida Cracker* Mountain Pleasure* Spanish Mustang/ Spanish Barb* Suffolk	Dartmoor Hackney Horse Lipizzan Rocky Mountain*	Clydesdale Friesian Gotland Shire
Pigs	Choctaw* Gloucestershire Old Spots Guinea Hog* Hereford* Large Black Mulefoot* Ossabaw Island (F)* Red Wattle* Saddleback	Tamworth	
Sheep	California Variegated Mutant/ Romeldale* Gulf Coast Native* Hog Island (F)* Santa Cruz (F)*	Cotswold Jacob (American)* Karakul (American)* Leicester Longwool Navajo-Churro* St. Croix* Tunis* Wiltshire Horn	Barbados Blackbelly* Dorset Horn Katahdin* Oxford Shetland

This summary is derived from a comprehensive North American livestock census conducted by the American Livestock Breeds Conservancy. The complete report of this census, *Taking Stock*, was published in 1994 by McDonald & Woodward Publishing Co. and is available from ALBC for $19.95.

* Unique to North America

(F) Feral population or breed of feral origin

Chapter 1

Why Rare Breeds?

What is a breed?

A simple and elegant definition of a "breed" has been suggested by Juliet Clutton-Brock: "a group of animals with a uniform appearance and behavior that distinguishes them from other groups of animals within the same species. When mated together, members of a breed reproduce this same type." That is, breeds breed true.

Each breed became consistent in appearance and behavior through a combination of human and natural selection. The diversity of environments, uses, aesthetic interests and practical needs created many distinct breeds, each of which fit a specific place. The rate of breed development varied, with some breeds existing in isolated areas for hundreds if not thousands of years, and other breeds created by breeders over a few decades. Taken together, the breeds within a species represent the species' adaptability and utility, providing the genetic breadth the species needs to survive and prosper.

AMERICAN MAMMOTH JACKSTOCK

Why are some breeds rare?

Throughout history, the mosaic of livestock breeds has always responded to changes in human needs and environmental conditions. Some breeds increased in popularity; others decreased. New breeds were formed from existing ones. Selection within breeds changed to meet market demands. Continuity was the rule, however, and a complete set of genetic choices in the form of breeds was passed from one generation to the next.

Things have changed in this century. Diversity in environments and management systems has been replaced by uniform systems of animal production. People across the country (and sometimes around the world) have begun to raise the very same breeds under the very same management. The use of climate-controlled confinement housing, increased veterinary support, high energy feed and other inputs has allowed breeders to ignore environmental adaptation and other survival qualities.

The goal of modern, industrial agriculture has been to maximize production, and the focus of selection has been the few most productive breeds within each species. Selection and management have sought continual improvements in performance from this narrowly selected genetic stock in its highly controlled environment.

Diversity as represented by pure breeds is neither valued nor sought, because it is assumed that the diversity present within industrial breeding programs is sufficient to meet any needs that arise in the future. The limited adaptability of industrial stocks, however, makes it unlikely that this assumption is true. Instead, genetic conservation with each species is necessary, and it should include a variety of breeds and types, especially those which have traits no longer found in industrial stocks.

Why conserve rare breeds?

The conservation of rare breeds protects the broad genetic base found in each livestock species. Thus, the adaptability and biological health of each species is maintained, ensuring that domestic animals can continue to thrive in a wide range of environments without elaborate and expensive support systems.

Rare breeds conservation also keeps livestock genetic resources in the hands of individual farmers and animal breeders. Livestock can be freely owned, used, and bred by farmers and breeders with few restrictions. In contrast, concentration of genetic ownership in the seed industry has resulted in the lack of access to diversity and the loss of many heirloom varieties.

Rare breeds are important historically and culturally. Like artwork, architecture, language, and other artifacts, rare breeds enlighten us about the interests, skills, and values of the people who preceded us. Unlike other elements of nature, these living animals also reflect our evolving relationship with the natural world. Rare breeds of domestic animals, as well as rare varieties of agricultural plants, represent the biodiversity which is closest to us and upon which we are most dependent. ❖

Chapter 2

Asses

ASSES, ALSO CALLED DONKEYS, are valuable though often underappreciated agricultural animals. Hardy, gregarious, and adaptable, asses are especially known for their braying voices and long ears. Their reputation for stubbornness results from a propensity to stand still while evaluating dangerous or unknown situations. This characteristic was acquired during the species' long evolution in rocky desert habitats in the Middle East and Northern Africa, which did not reward the flight response found in their horse, zebra, and Asiatic ass cousins, which are adapted to life on the plains.

Asses were domesticated from several wild subspecies in the Mediterranean region. They proved useful for draft and transportation. Asses were also used for meat and occasionally for the production of milk. The milk is known for its cosmetic properties and still used today in some parts of the world.

The ass species has undergone less elaborate breed development than have the other species, resulting in relatively few standardized breeds. Landraces, shaped chiefly by geographic isolation, are more common. Asses are generally grouped into one of three types, defined by their size and uses.

The largest breeds of asses are the size of horses, standing 14 hands (56") or taller at the withers. These breeds have been developed in many regions of the world, and they have been selected primarily for the production of mules – the sterile offspring of male donkeys and female horses. Mules are superb work animals, combining the hardiness of donkeys with the size and strength of horses. Mule breeding has gone on since antiquity in several parts of the world, especially in Spain, France, Italy, the Middle East, Central Asia, and China.

Mule breeding has also been an important enterprise in the history of the United States. Mules, along with asses, were first imported from Spain in 1520. The American Mammoth Jackstock, this country's primary mule-producing ass, was developed beginning in the late 1700s. The breed was prized, and its economic value often approached that of cattle and horses.

POITOU ASSES

Mechanization of agriculture in this century caused the market for mules to collapse, and the Mammoth Jack fell into decline, as did all of the large ass breeds around the world. Where large breeds of asses survive, they are rare and should be considered high conservation priorities.

Medium size asses, often called donkeys, number in the millions globally. Animals in this group stand about 9–14 hands (36–56") high at the withers. They are used chiefly for draft, packing, transportation, and other types of work, as well as for the production of medium size mules. Only a few distinct breeds have been developed throughout the world, although some local strains have been highly selected for specific tasks and thus developed the genetic consistency expected of a breed. The elite white riding donkeys from Muscat in the Arabian peninsula are an example. North America has two medium size breeds, the Standard and the Large Standard.

The smallest donkeys are the miniatures, which stand less than 9 hands (36") high at the withers. This group includes animals historically selected for working under difficult conditions, including those found on islands where large size would be a disadvantage. The best known population of miniature donkeys evolved in the Mediterranean region. This breed has now become numerous in North America.

Ass breed resources in America descend almost entirely from Spanish stocks, though there have also been contributions from other European countries. The earliest imports were of medium size Spanish asses used as pack animals. These donkeys became the basis for feral populations in the Southwest, called "burros." During the late 1700s and early 1800s, larger ass breeds were imported from Spain and other European countries to be used in the development of the American Mammoth Jackstock breed. Miniature strains from the Mediterranean were imported as companion animals beginning in 1929.

The American Donkey and Mule Society (ADMS), founded in 1967, has been instrumental in the promotion of ass breeds and breeding in the United States. The ADMS holds registries for all donkey and mule breeds and provides information on issues of selection, breeding, and management.

American Mammoth Jackstock

Origin: United States
Classification: Standardized breed
Distribution: United States, with a small population in Canada
Status: Rare

Most donkeys bray. Mammoth Jacks roar, and when a barn full of Jacks has been set off, it's an impressive event! This breed was developed beginning in the earliest days of the United States, and it has been an integral part of American agricultural history.

George Washington was the father of the nation, but he was also one of the leading agricultural innovators of his day. Among his many interests was the improvement of livestock, including the development of an American ass breed which could be used to produce strong work mules. Washington, with Henry Clay and others, obtained by gift and purchase a small number of jacks and jennies of the finest European breeds in the late 1700s and early 1800s. Interest increased quickly. Washington, for example, was offering his jacks for stud service by 1788. The Catalonian ass from Spain was of primary interest to American breeders, but the Andalusian (from Spain), Maltese (from Malta), Poitou (from France), Majorcan (from Majorca), and Italian strains were also used. Breeders blended these strains together, selecting the offspring for size, soundness, and strength. In this way, they created the American Mammoth Jackstock breed.

Large breeds of asses were found in Kentucky by 1800, and Kentucky, Tennessee, and Missouri became the primary mule-producing states. Between 1830 and 1890, several thousand large asses were imported from Spain and other parts of Europe, broadening the genetic base of the breed. The substantial investment in these imports demonstrates the significance of the mule industry at that time. A registry was established in 1888, and a second registry in 1908. These merged in 1923 as the Standard Jack and Jennet Registry (SJJR), which continues today under the name of the American Mammoth Jackstock Registry. The population of American Mammoth Jackstock peaked in 1920, with an estimated five million animals in the national herd.

Mammoth Jacks are sturdy and tall, with massive legs and large, well-made heads. The ears are especially long, often measuring 33" from tip to tip. Selection has always been made for size and substance. Traditionally, males were expected to stand at least 14.2 hands (58") high at the withers and females 14 hands (56"), and many animals were a good bit taller than this. The weight varied with the height and ranged between 900 and 1,200 pounds.

During the 1950s, in response to the dramatic decline in the number of American Mammoth Jackstock, the Standard Jack and Jennet Registry (SJJR) lowered the breed's height requirements to 14 hands (56") for males and 13.2 (54") for females. When the American Donkey and Mule Society (ADMS) established its registry for American Mammoth Jackstock in 1967, it followed the SJJR breed standard. Ironically, in the 1980s, when breed numbers were again increasing, the SJJR raised its height standard back to the original criteria. This is the origin of the difference in breed standards between the two registries. The genetic effect of the lowering of the height standard is not documented, though it is true that animals which would not have been tall enough historically have been admitted into the breed in the past 40 years. Better documentation of today's population would be useful in understanding the effect of these policy changes on the breed's genetic status and conservation needs.

Black used to be considered the only suitable color for the breed, as black Jackstock bred to Percheron mares produced dark colored mules

American Mammoth Jackstock

that were easy to match as teams. For the last few decades, however, the market has favored sorrel draft mules, produced by breeding a sorrel Jack and a Belgian mare. As a result, the predominant color of the American Mammoth Jackstock has also become sorrel. With this change in color has come a change in type as well, as the sorrel animals tend to be more coarse in conformation than the blacks.

The primary function of American Mammoth Jackstock has always been to produce draft mules. Today, draft mules are as likely to be used in recreation as in agricultural work, and their numbers are increasing. The market for riding mules is also increasing, with gaited mules in a variety of colors especially in demand. This trend may or may not benefit the American Mammoth Jackstock breed, because Large Standard donkeys, which are lighter boned and more refined, are often used to produce these riding mules.

The American Mammoth Jackstock is found almost exclusively in the United States, with a small population in Canada. There are 3–4,000 animals alive today, with the largest numbers in Kentucky, Tennessee, Missouri, and Texas. Only a few hundred of the historic type black Mammoth Jacks remain, however, and conservation of these strains is a priority for the breed.

Poitou

Origin: France (Poitou)
Classification: Standardized breed
Distribution: France, with small populations in a few other countries
Status: Critical

The Poitou ass is an ancient French breed valued for the production of mules. The breed was developed in the Poitou region of western France, where mule breeding has been documented for over one thousand years. Standardization of the Poitou breed occurred by about 1700, and the modern studbook was established in 1884. In France the breed is referred to as *Baudet Poitou*, as *baudet* means "sire of mules," distinct from the term *ane* for common donkey.

The French mule breeding industry was a highly developed and lucrative enterprise. Poitou jacks were bred to mares of a specific horse breed – the Mulassier Poitevin – to produce mules for both riding and work. While mules were promoted across Europe, breeding stock was closely held in the Poitou region. The breed nearly became extinct after World War II, when the demand for mules collapsed. The Poitou's limited geographic area increased its vulnerability. Fewer than 80 Poitous survived in 1980.

Recent conservation efforts in France have been organized to protect the breed from extinction, and several organizations have been involved. The Parc Naturel Regional du Marais Poitevin has made conservation of the Poitou and the Mulassier Poitevin a part of its mission to protect the domestic heritage as well as the wild environment. La Sabaud, a network of breeders, exists to promote the breed and raise funds for its conservation. La Sabaud support the efforts at the Asinerie Nationale Experimentale, an experimental farm that breeds Poitous in conjunction with the Parc. L'Association des Eleveurs des races Equine, Mulassiere et Asine, Baudet du Poitou is the registry organization.

These conservation efforts are getting results. The breed has increased in numbers, and today there are about 150–200 Poitous registered in the studbook. It is thought, however, that some of these animals may carry falsified papers, having been sold by speculators at high prices because of the breed's rare status. A challenge facing the breed is to continue to inventory and document the asses claimed to be Poitous.

The North American Baudet du Poitou Society has been formed to promote stewardship of the Poitou in the United States and Canada. The Society has inventoried the registered animals in the United States and serves as the liaison between North American breeders and the French

Poitou Ass

organizations. There are about 30 Poitous in North America.

Poitous have the conformation and strength typical of mule-breeding asses everywhere. They are tall, standing 14–15 hands (56–60") high at the withers and weighing 750–950 pounds. Their heads are large and long, with long ears, and their bodies are muscular with heavy bone. The feet should be large. The most striking feature of the breed is its long hair, which may hang in cords from the entire body. Poitous are found in black or dark brown, without a dorsal stripe. Other colors are not accepted. ❖

Chapter 3

Cattle

CATTLE ARE POSSIBLY the most revered of the livestock species around the world. Large and imposing, cattle have nevertheless adapted to an incredible range of natural environments and human societies, providing people with milk, meat, leather, and draft power. In many cultures, cattle also have spiritual, economic or political importance far beyond the monetary value of the animals themselves. As a response to these many uses and habitats of cattle, a wide array of breeds have been developed.

Spanish explorers first brought cattle to the Americas beginning in the early 1500s. Spanish cattle rapidly populated southern North America, Central America, and South America. These cattle were hardy and rugged, and they adapted readily to the new environments. Differences in natural and human selection across the Americas led to the evolution of several breeds. These make up a breed family called *criollo* cattle; the term *criollo* means "of European origin but born in the New World." North American criollo breeds include the Corriente, Florida Cracker, Pineywoods, and Texas Longhorn.

DUTCH BELTED CATTLE

Cattle from England and Northern Europe were also imported to North America beginning in the early 1600s. Three of these early importations were significant. Cattle from Normandy and Brittany were brought to Quebec, where they became the Canadienne breed. North Devon cattle from England were brought to New England, where they were known as Milking or Red Devons. Channel Islands cattle were also imported to New England, where they evolved into historic landrace populations of which only the Randall Lineback remains.

These cattle – Spanish breeds in the South and English and European types in the North – served a variety of subsistence niches in America for over 200 years. About 1800, a more deliberate introduction of cattle breeds began. During the century that followed, several improved cattle breeds were imported, including the Aberdeen Angus and Ayrshire from Scotland, the Hereford and Shorthorn from England; Channel Islands cattle from France; and Holstein and Dutch Belted cattle from the Netherlands. Of these breeds, the

Shorthorn (also known as the Durham) was by far the most valuable. People needed multi-purpose cattle, and the Shorthorn combined excellent dairy and beef qualities as well as the size and strength necessary for use as oxen. It soon became the most popular breed in America.

By 1900 the market had shifted to favor the use of specialized beef and dairy breeds. The Hereford and Angus came to dominate the beef industry, while the Ayrshire, Jersey, and Guernsey were the most numerous of the dairy breeds.

Imports since 1900 have further increased the diversity of cattle breeds in the United States, especially those breeds used for beef. Zebu cattle, originally from India, rapidly became popular in the southern U.S. because of their tolerance of hot climates. Zebus were the foundation of the American Brahman and Santa Gertrudis breeds. African genetics, in the form of N'dama and Afrikander cattle, have been used in the development of the Senepol and Barzona breeds, also known for heat tolerance. Processors' interest in large framed cattle was answered through importation of the Charolais, Chianina, Simmental and other "continental" breeds, named for their place of origin on the European continent.

The large number of beef cattle breeds – and the genetic diversity they represent – has been a cornerstone of success for the beef industry, allowing producers to respond to changing market demands. Yet diversity has been conserved not formally but informally, because of the broad range of habitats in which beef cattle are raised, the accessibility of markets, and decentralized approaches to selection. It is because of this informal conservation process that farmers and breeders will have the diversity they need for new production and market niches, such as grass-fed beef.

The dairy industry presents a sharp contrast, as it rests almost entirely on the use of a single breed, the Holstein. The Holstein is known for its adaptation to confinement dairying, and the cows produce more milk under such conditions than do those of any other breed. As a result, it has prospered at the expense of all other breeds in the past fifty years. The success of the Holstein, however, rests on the availability of high levels of inputs, including large amounts of grain and veterinary support.

The resurgence of lower cost, grass-based dairying as a production niche has caused many dairy farmers to rethink the industry's reliance on the Holstein. Grass-based production requires cows that are active grazers, able to maintain body condition, produce milk, and reproduce efficiently on a forage diet. Farmers looking for these qualities have turned to the Ayrshire, Brown Swiss, Jersey, and other "colored" dairy breeds. These breeds are experiencing a revival of interest and are now being used as purebreds and for crossing on Holsteins to improve cow fertility and grazing ability in the offspring.

The pressures of economic consolidation and vertical integration, significant in the swine and poultry industries, have had less obvious impact on cattle. Nonetheless, there is increasing consolidation among the companies that buy milk and beef from farmers. This process is gradually having two negative effects: the overall lowering of prices paid and the further discounting of animals which do not conform to a standard industrial type. The cattle industry, built upon a foundation of genetic diversity, cannot afford to let short term market pressures eliminate rare breeds and thus the diversity essential to its future success.

Ankole-Watusi

Origin: East Africa
Classification: Landrace in Africa, Standardized breed elsewhere
Distribution: East Africa, United States, with small populations in Australia and Europe
Status: Rare

Ankole-Watusi cattle, with their long, large horns, are among the most striking members of the bovine kingdom. Herds resemble slow moving, multi-colored forests of bare trees as their horns sway with every step.

The Ankole-Watusi breed is part of the Sanga family of African cattle breeds, which originated over 2,000 years ago from a combination of the Egyptian Longhorn cattle of Africa and Zebu Longhorn cattle originally from India. Sanga cattle spread throughout eastern Africa, and many distinct breeds evolved.

Cattle have traditionally been valued in Africa as ceremonial animals and as symbols of wealth and power. Ownership of cattle established one's position in society, and the beauty of one's herd (especially the shape and size of the animals' horns) was significant. Natural selection also had a role to play in creating cattle that were hardy, efficient grazers able to thrive on rough forage. The horns are part of adaptation to a hot climate by allowing dispersal of excess body heat. This unusual combination of selection pressures has resulted in a resilient breed of great beauty. Particularly remarkable are the cattle of Uganda, Rwanda, and Burundi. In Uganda, the Nkole tribe's variety is called the Ankole, while in Rwanda and Burundi, the Tutsi tribe's variety is called the Watusi. When the cattle were exported from the region to Europe and later to America, these strains were combined.

The Ankole-Watusi is medium in size and elegant in appearance. Cows weigh 900–1,200 pounds and bulls 1,000–1,600 pounds. The most noticeable feature of the breed is its horns, which have a large base and great length. Crescent or lyre-shaped horns are the most desired in Africa, but American breeders have also selected for a

Ankole-Watusi Cow

lateral horn conformation that increases the length from tip to tip. Cattle may be solid colored or spotted. Dark red predominates, although dun and black also occur. Some unusual spotting patterns are found in the breed, including one in which color runs across the top of the animal with white on the lower body. Many types of speckling are also seen.

Unlike the purposeful importations of many breeds, the path of the Ankole-Watusi to the United States was indirect if not accidental. The cattle were initially exported from East Africa to Europe during the early 1900s for use as exhibition animals in zoos. U.S. zoos and individual breeders then imported Ankole-Watusi cattle from Europe during the 1920s and 1930s. In the last few decades, most zoos have dispersed their collections, putting the majority of the cattle in the hands of private breeders.

The Ankole-Watusi International Registry was formed in 1983 to promote the breed in North America. Founding members shared a commitment to the Ankole-Watusi, though they have had different priorities for the breed. Some have emphasized its exotic appearance and others its commercial qualities. Though often dismissed as a novelty, the Ankole-Watusi has much to offer the U.S. beef industry. Adaptation to harsh environments, excellent maternal abilities, high butterfat milk, and lean beef are among its assets.

Approximately 1,500 Ankole-Watusi cattle are found outside of Africa, and about 80% of these are in the United States. The population of cattle in the breed's home region of Rwanda, Burundi, and Uganda has been numerous, though it is now in dramatic decline due to civil war, economic upheaval, and government-supported cross-breeding schemes. This makes stewardship of the American population an essential element in the Ankole-Watusi's global survival.

Ayrshire Cow

Ayrshire

Origin: Scotland
Classification: Standardized breed
Distribution: International
Status: Watch

The Ayrshire is a dairy breed named for its home region, County Ayr in southwestern Scotland. Unlike many other breeds which trace their origins to landrace populations, the Ayrshire was purposefully developed as a standardized breed. Breed development began in the mid-1700s, when native black and black and white cattle were crossed with Flemish and Teeswater cattle and with cattle from the Channel Islands. The goal was to increase milk and meat production. The breed was recognized in 1814, and selection during the 1800s further improved dairy qualities, with emphasis on milk for the manufacture of cheese and butter. A herdbook was established in 1877.

Ayrshires were first imported to North America in the 1830s, but the majority of imports came after 1850 and thus reflected more improved types. The breed became especially popular in New England, and Daniel Webster was among its early proponents. Ayrshires also became widespread in eastern Canada. The American Ayrshire Breeders Association was formed in 1895, and the breed's numbers grew dramatically after 1900.

Ayrshires have alert and active dispositions, and the breed is considered the most stylish of the dairy cattle. Its handsome, lyre-shaped horns are distinctive. Cattle are white with dark red or reddish brown spots and speckles. Some cattle have brown legs, as if they have their work boots on. Ayrshires are medium in size; cows average 1,100 pounds and bulls 1,600 pounds.

In spite of their bovine beauty, Ayrshires are best known as hardy, useful cattle, well adapted to cold climates and rugged environments. The cows are excellent grazers, able to maintain condition, reproduce efficiently, and produce up to 20,000 pounds of milk per year, primarily on a forage diet. The cows' longevity adds to their value.

Over the last 30 years, the number of registered Ayrshires in North America has declined significantly. The breed cannot compete with the Holstein in milk quantity produced under confinement conditions. On the other hand, the Ayrshire thrives under low-input, grass-based dairying, and the growth of this production niche has increased interest in the breed. Ayrshire numbers are relatively stable and numerous globally, though the declining status of the North American population has attracted conservation efforts.

Belted Galloway

Origin: Scotland
Classification: Standardized breed
Distribution: International
Status: Watch

Galloway cattle originated in the rugged hill country of southwestern Scotland. The native cattle of this region were consistently hardy and thrifty, though they varied in color, conformation, and horn type. During the late 1700s and early 1800s, the Galloway breed was developed through selection for more standardized, polled beef cattle that retained environmental adaptation. Galloways came to be celebrated for their efficiency on rough forage, maternal ability, and high quality beef. A herdbook was established in 1877.

Most Galloway cattle are black, but several other colors are also found in the breed. The most distinctive is black with a white belt. A separate herd book for belted cattle was established in Scotland in 1928, and this division of the strains by color has led to the formation of the Belted Galloway breed. This breed remains very similar to the Galloway, however, and shares most of its other characteristics.

The origin of the belted color pattern is not clear. This pattern occurs in several breeds found in different places, most notably in the Lakenvelder breed in the Netherlands (called the Dutch Belted in North America). The belted pattern has

Belted Galloway Cattle

been selected against and has disappeared in some breeds, such as the Welsh Black and the Kerry. In the Belted Galloway, however, the belted pattern has been fostered as the standard for all animals of the breed.

Beneath the flashy appearance is found a very useful animal. The Belted Galloway is well adapted to cold, rugged climates; as part of this adaptation, the cattle grow a shaggy hair coat in the winter. In the summer, Belted Galloways shed this coat and tolerate warm climates better than do most other cold-adapted cattle. Galloways are medium in size, with cows weighing about 1,000 pounds and bulls about 1,600 pounds. Cows show strong maternal instinct, including the ability to protect calves against predators.

Belted Galloway beef, like that from Galloways and other Scottish breeds, is of high quality and can be marketed in a variety of specialty niches, including grass-fed and organic beef. The long body conformation increases the production of high priced cuts of beef. Hides may be used to make rugs, and the belting pattern combined with the shaggy hair makes these products distinctive and attractive.

Belted Galloways were imported to North America beginning in the 1940s, and the Belted Galloway Society in the United States was founded in 1951. The breed is increasing in numbers in North America and globally. In Europe the breed is considered a "light grazer," meaning that its relatively light body weight makes it preferred for grazing where heavier animals would damage marginal lands. Belted Galloways are also known for browsing ability, giving them added value in the management and diversification of grasslands.

Canadienne

Origin: Canada
Classification: Standardized breed
Distribution: Canada, with a small population in the United States
Status: Critical

The Canadienne is one of the few historic breeds of livestock developed in North America. It descends from cattle of Normandy and Brittany which were brought to Canada between 1601 and 1660. This population was shaped by natural selection in the rugged environment of eastern Canada, and the resulting breed became known as the Canadienne. The breed is closely associated with Quebec, and even today the majority of the cattle are found in this province.

Canadiennes are black, brown, tan, or russet with a pale fawn muzzle and udder. Many have a lighter colored stripe along the back. The black-tipped horns curve up and turn back toward each other. Cows weigh about 1,100 pounds, and bulls about 1,600 pounds. Milk production averages about 15,000 pounds over 305 days, with about 4.35% butterfat and 3.7% protein. Most significant is the breed's ability to produce milk on poor forage and under very challenging conditions. No other dairy breed demonstrates this combination of hardiness and productivity.

The Canadienne was the dominant dairy type in eastern Canada until the importation of improved European breeds in the early 1800s. In 1850, the Canadian government officially discouraged the use of the Canadienne in favor of improved breeds. This unfortunate policy was reversed in 1883, when the breed was again recognized as valuable for its superior environmental adaptation. Pure breeding was again encouraged, and a herd book was established in 1886.

The breed flourished for several decades, though became rare again during this century. The dairy industry's emphasis on quantity of production under confinement husbandry left the Canadienne unable to compete with the Holstein. Brown Swiss genetics were introduced as a way to increase milk production, though the result was to dilute the characteristics of the Canadienne

Canadienne Cow

DEVON CATTLE

without gaining any market advantage for the breed. Fortunately, pedigrees were documented and some cows continued to be bred to purebred bulls, so that pure stocks continue to exist as well as those with Swiss breeding. Now this policy has been reversed and farmers are again choosing the traditional cattle that have served them so well.

The Canadienne breed is critically rare, with fewer than 300 cattle in North America today, most of them in Quebec. The breed merits attention for use in grass-based dairying, especially where the quality of forage is low. Under these conditions, the Canadienne will be able to demonstrate its value to dairy farmers and thus gain a broader base of support.

Devon

Origin: England
Classification: Standardized breed
Distribution: United States, with related populations in several other countries
Status: Critical

Devon cattle were developed over several centuries on the southwestern peninsula of England. The breed's name comes from Devonshire, though cattle were also raised in the neighboring counties of Cornwall, Somerset, and Dorset. Devons were valued for the production of both high quality beef and the rich milk used in Devonshire cream. They were especially well regarded as the quickest and most active oxen in the British Isles. The breed also had the reputation as an easy keeper, able to thrive on rough forage.

The Pilgrims brought Devon cattle with them to New England beginning in 1623. The hardiness and practicality of the breed, plus the availability of Devon cattle near the ports of departure, made Devons an obvious choice for early immigrants to the Americas. The breed became well established in New England during the 1600s and spread down the coast as far as Florida during the 1700s and 1800s. The cattle also went west, as Devon oxen were the draft animals of choice on the Oregon Trail. A registry was established for the breed, and herdbooks have been published since 1855.

By the late 1800s, however, the Devon had been gradually replaced by the Shorthorn, a more productive multi-purpose breed. By 1900, Devons were rarely seen outside of New England. Only in this region did the breed continue to be valued for hardiness and the ability to thrive under rugged conditions, qualities in which it surpassed the Shorthorn.

By the 1950s, the market for multi-purpose cattle had disappeared and the Devon breed neared extinction. In response to this challenge, the breeder community split. The majority of breeders began to select intensively for beef production in order to compete with other beef breeds. This population was called the Devon or the Beef Devon. A few breeders continued to select their animals for meat, milk, and draft; this population was called the American Milking Devon. Beef-selected Devon cattle populations are found in several other countries, including Britain and Brazil, though the long separation of these populations makes it unclear whether they are still a single breed. Research continues into the status and characteristics of Devons internationally.

Devon cattle are ruby red with black-tipped, creamy white horns. Cattle are medium in size, with cows averaging 1,100 pounds and bulls 1,600–2,000 pounds. The appearance is compact with a straight topline, square set legs, and well-formed udders.

The Devon has several assets which recommend it for use in today's competitive beef market. The breed's history as a dual-purpose animal gives it greater maternal ability than most other beef breeds. (Any decline in this characteristic is easily corrected through the use of a Milking Devon bull, which can be registered in the Devon herdbook.) The Devon is known for high quality beef, and the breed's hardiness and grazing ability makes it an excellent choice for grass-based production. The Devon is distantly related to other British and European breeds found in North America, so it provides great hybrid vigor when used in commercial crossbreeding.

The Devon is sometimes called the North Devon to distinguish it from the South Devon, a large, dun beef breed to which it is not closely related.

Dexter

Origin: Ireland
Classification: Standardized breed
Distribution: Australia, Britain, Canada, Ireland, United States
Status: Rare

Dexter cattle are among the smallest of cattle breeds in the world, standing 40" tall and weighing 700–900 pounds. Though size is the breed's most distinguishing characteristic, the Dexter is a useful and productive multi-purpose animal.

The Dexter originated in southern Ireland during the early 1800s. It was developed from the Kerry, an Irish dairy breed, through selection for smaller size and improved beef qualities. The breed name came from a "Mr. Dexter," who promoted the cattle during the mid-1800s. The Dexter became popular with smallholders in Ireland and in England, who appreciated its efficiency in producing both milk and beef on limited acreage. Dexters were imported to North America beginning in 1910.

Even after the Dexter gained an identity as a breed, its history remained intertwined with that

DEXTER CATTLE

Dexter and Milking Shorthorn Cows

had more dairy selection produce more milk, while those which have been selected for beef produce less. The milk produced is high in solids, making it ideal for butter and cheese production. Dexter beef is lean and high in quality. The small size of the carcass makes the breed an excellent choice for use in direct marketing programs. Dexters are good browsers and can rid pastures of some pest plants, and they may also be used as oxen.

Dexter cattle are increasing in numbers in North America and globally, and the breed seems destined to succeed. The challenge facing breeders, however, is to maintain historic selection practices so that the Dexter's production qualities are conserved and promoted.

of the Kerry. For many years the two breeds were registered in a single herdbook, and some people considered Dexters to be a dwarf type of Kerry. Recent bloodtyping research, however, has determined that the Dexter and the Kerry, though closely related, are genetically distinct breeds. They should not be crossbred with each other.

Dexter cattle are solid and compact in appearance. Most Dexters are black, though red and dun are also found. The cattle are horned, and their black-tipped white horns arc upward. Two body conformations are found within the breed: cattle with normal bodies and very short legs, and cattle which are proportionately small in every dimension. Because the short-legged type occasionally produces nonviable offspring, it has fallen out of favor, while the proportionate type has become more popular.

The Dexter has always attracted attention because of its size, and it has sometimes been marketed as a novelty or ornamental breed. This practice has obscured the breed's production value. Dexters are hardy, forage-efficient cattle with excellent maternal qualities. As with other dual-purpose breeds, the quantity of milk produced varies between strains; those strains which have

Dutch Belted

Origin: The Netherlands
Classification: Standardized breed
Distribution: The Netherlands, United States
Status: Critical

The Dutch Belted is a dairy breed named for its country of origin and its unusual color pattern: black with a bright white belt around its middle. In the Netherlands, it is also known as the Lakenvelder. The term *laken* means a sheet or blanket around the body, and the name is shared with the Lakenvelder chicken breed, which has a similar black and white pattern.

The Dutch Belted is a historic breed, known in the Netherlands since the 1500s and standardized since the 1700s. It was selected as a specialized dairy breed able to convert lush pastures into milk. The Dutch Belted is related to the Holstein (or Friesian), though the two breeds have long diverged.

Dutch Belted cattle were imported into the United States between 1840 and 1900. P.T. Barnum was one of the first importers, and he included the flashy cattle in his circus shows. The Dutch Belted Cattle Association of America (DBCAA) herdbook was organized in 1886. While never very numerous in the United States, the breed was well known and nationally distributed, managing to carve out its own niche in the dairy industry. The Dutch Belted became rare this century, however, when the dairy industry began to focus exclusively on Holsteins as the most productive cows for confinement systems.

By the 1970s, the Dutch Belted breed neared global extinction. The DBCAA lapsed into inactivity, and only a few purebred breeders remained. Unfortunately this decline coincided with the U.S. Department of Agriculture's dairy buyout program of the 1980s, when thousands of dairy cattle were sold for beef as part of the national dairy herd reduction designed to raise milk prices. Many Dutch Belted cattle were lost through this program. It was only through the action of a handful of breeders, especially Kenneth and Winifred Hoffman of Earlville, Illinois, that the Dutch Belted breed survived in the United States. Their work had a global impact as well, since semen from the Hoffmans' bulls has been used in the Netherlands.

In recent years, there has been a revival of interest in the Dutch Belted breed, chiefly among farmers interested in grass-based dairying. Dutch Belted cattle are an excellent fit for this niche. They are medium in size, with cows weighing 1,200 pounds and bulls 1,700 pounds. The cows easily maintain condition and produce milk on grass, especially where the forage is of good quality. They consistently demonstrate reproductive efficiency and longevity of production. Dutch Belted bulls are prepotent, stamping even crossbred offspring with the breed's qualities. For this reason, Dutch Belteds are being used as purebreds and for crossing onto purebred and grade Holsteins to increase the health, fertility, and grazing ability of commercial herds.

The American Livestock Breeds Conservancy (ALBC) began management of the Dutch Belted Cattle Association of America's registry in

Dutch Belted Cow

1993. ALBC facilitated the Association's reorganization, providing the structure and representation necessary for the breed to survive. The DBCAA registry has established grade-up and recovery programs to increase numbers in the breed. Fewer than 1,000 Dutch Belted cattle exist today, including about 200 documented purebred cows in the United States.

Dutch Belted cattle are black (or occasionally red) with a white belt. The belted pattern is dominant, so that crossbred Dutch Belted cattle (especially Dutch Belted-Holstein crosses) may have the markings of a purebred. For this reason, crossbred cattle have sometimes been marketed as purebreds. This, plus the number of undocumented, unregistered purebred cattle, has been an obstacle to conservation of the breed. Nevertheless, the Dutch Belted is in a stronger position now than it has been in several decades. If the breed can take advantage of the new opportunities in grass-based dairying, and if interest in use of the breed can be translated into purebred breeding, the Dutch Belted may see its best days ahead.

Florida Cracker

Origin: United States (Southeast)
Classification: Landrace
Distribution: United States (Southeast, especially Florida)
Status: Critical

The Florida Cracker is one of the oldest breeds of cattle in the United States, descending from Spanish cattle brought to the New World beginning in the early 1500s. As the Spanish colonized Florida and other parts of the Americas, they established low-input, extensive cattle ranging systems typical of Spanish ranching. The Florida Cracker and other breeds which developed under these conditions are called *criollo* cattle, which means "of European origin but born in the New World."

The Florida Cracker breed was shaped primarily by natural selection in an environment that is generally hostile to cattle. This has resulted in a breed that is heat-tolerant, long-lived, resistant to parasites and diseases, and productive on

FLORIDA CRACKER CATTLE

the low quality forage found on the grasslands and in the swamps of the deep South. It was not until the importation of Zebus from India and the development of the American Brahman breed this century that the Florida Cracker had competition from other heat-tolerant cattle. Not long afterwards, the development of parasiticides and other medications allowed British and European breeds to survive in the deep South, and thus the Florida cattle industry was further diversified.

This influx of new breeds very nearly caused the extinction of the Florida Cracker breed. By the mid-1900s, the majority of purebred cows had been crossbred, first to Brahmas and then to British and European breeds. The Cracker cow provided the maternal ability and hardiness necessary for crossbreeding programs to succeed, and the genetic distance between the Florida Cracker and other breeds created exceptional hybrid vigor in the offspring. The credit for success, however, was always given to the improved breeds, and the Florida Cracker was largely abandoned. It was only through the efforts of a few Florida families, stubbornly resistant to "progress," that the breed survived in its pure form.

As one would expect of cattle adapted to the heat, Florida Cracker cattle are small, with cows weighing 600–800 pounds and bulls 800–1,200 pounds. They exhibit the angular conformation typical of Spanish cattle adapted to harsh conditions. Horn style and shape vary, including very long and twisted horns as well as smaller, more crumpled shapes. Polled cattle are also found. Dwarf types of Florida Crackers, called "Guineas," occurred historically and were well regarded as milk producers.

Colors in the breed vary widely, including almost all of the colors known in cattle. This is a legacy from the breed's ancestors in the South of Spain, where even today ranchers prefer multicolored herds. In Florida, breeders also selected for a variety of colors and patterns as a way to aid recognition of their cattle on range. Generally, solid red, dun, black, and brindle colors predominated in south Florida herds, while in the north, herds were more likely to contain spotted and roan animals, some with distinctive color-sided and linebacked patterns.

The state of Florida has been a leader in the conservation and promotion of the Florida Cracker breed during the past two decades. Cracker cattle are considered a living part of Florida history, and herds have been maintained at several state parks and forests. The state has supported the establishment of the Florida Cracker Cattle Association and a breed registry, which is operated by the American Livestock Breeds Conservancy. The Florida Cracker cattle breed is still quite rare, with fewer than 500 animals alive today, but its prospects are brighter than they have been in a long time.

Galloway

Origin: Scotland
Classification: Standardized breed
Distribution: International
Status: Watch

The Galloway breed originated in the rugged hill country of southwestern Scotland. The native cattle of this region were consistently hardy and thrifty, though they varied in color, conformation, and horn type. During the late 1700s and early 1800s, the Galloway breed was developed through selection for standardized, polled beef cattle that retained environmental adaptation. Galloway cattle became a valuable element of the trade between Scotland and England.

The Galloway is related to the Aberdeen Angus, which was developed in northeastern Scotland about the same time. A herdbook for the two black, polled beef breeds was established in 1862, with a separate Galloway herdbook opened in 1877. Since then, the breeds have diverged, with the Angus selected for rapid growth on better feed, and the Galloway for its ability to thrive on poor forage in a cold, wet climate.

Galloway cattle were imported to the United States beginning in the 1850s, and the American Galloway Breeders Association was founded in

1882. The Galloway's success initially rivaled that of the Angus, though by 1900, it was found in a smaller niche as a range breed on the Plains.

Galloway cattle are polled and medium in size. Cows weigh about 1,200 pounds and bulls 1,800–2,000 pounds. The long body conformation increases the production of high priced cuts of meat. Galloway beef is of great quality and can be marketed in a variety of specialty niches, such as grass-fed and organic beef. The cattle grow a long, thick coat in the winter as part of their adaptation to the cold. Galloway hides were substitutes for buffalo robes in the early 1900s, and they can also be made into rugs.

Black is the most common color in the breed, with red and dun also found. White Galloways occur more rarely. They are white with black or red points, including ears, eye rings, nose, feet, and teats. Within herds of white Galloways it is not unusual to find roan and speckled animals as well. Belted Galloways, black with white belts, also originated within the Galloway breed, though are generally registered with a separate association. Some belted animals continue to be registered in the Galloway herdbook, and the color is recognized as part of the Galloway breed.

The Galloway is rare in North America, but it is increasing in numbers globally with an estimated population of about 10,000. In Europe the breed is considered a "light grazer," meaning that its relatively light body weight makes it preferred for grazing where heavier animals would damage marginal lands. Galloways are also known for browsing ability, giving them added value in the management and diversification of grasslands.

In North America, the Galloway stands out for its forage efficiency, hardiness, maternal qualities, and excellence of beef. The breed's history as a pure breed means that Galloway cattle are genetically consistent and impart outstanding vigor to crossbred offspring. Black offspring have the advantage of receiving the same premium price given to Angus cattle as part of the Certified Angus Beef program. Recent market changes favoring efficient production and high quality beef may lead to a revival of interest in the Galloway, a breed which is far too useful to remain rare for long.

Galloway Cow

GUERNSEY CATTLE

Guernsey

Origin: England (Channel Islands)
Classification: Standardized breed
Distribution: International
Status: Watch

The Guernsey cattle breed originated in the Channel Islands between France and England and is named for the Isle of Guernsey, which is the most western of the group. The Jersey breed was developed on the neighboring island of Jersey, and the two breeds are related.

Cattle were first brought to Guernsey from Brittany over 1,000 years ago. Subsequent imports brought cattle from Normandy, and this founding population was variable in size and in color, including white, red, black, brindle and fawn animals. Over the next several centuries, cattle on the island were selected for the richness and quantity of their milk, and performance was gradually improved.

Channel Islands cattle were exported to England beginning in the 1700s, where they were widely used in the establishment and improvement of other breeds. There was extensive trade between the Channel Islands and England, France, and the Netherlands, and cattle were part of this trade until the early 1800s. At that time, imports into Guernsey were stopped and the island herds were closed. Since then, the export of breeding stock from the islands has been a major agricultural enterprise.

The closing of the island herds encouraged the standardization of the islands' breeds. Cattle shows, which began in 1828 on Guernsey, hastened the process. Guernsey breeders selected their cattle for consistency of color within a range of fawn to golden with white spotting and for golden skin. Today, Guernseys are medium to large in size, with cows weighing 1,400 pounds and bulls 2,000 pounds. The breed is usually horned, though polled strains have been developed. Cows are noted for their quiet dispositions.

A distinctive characteristic of the breed is the golden color of its milk, which results from exceptionally high levels of carotene, a precursor to Vitamin A. It is thought that the Guernsey excels in its ability to absorb this nutrient and transfer it to butterfat. Guernsey milk has been promoted under the trademark "Golden Guernsey." Butter made from the milk is also distinctively golden.

Guernsey cattle were first imported to North America in the 1830s, with importations peaking in the late 1800s and early 1900s. The Guernsey Cattle Club was formed in 1877. The breed was a significant genetic foundation for the American dairy industry, contributing to the general mix of dairy cattle as well as being used as a purebred.

Guernsey cattle have had a prominent place in American agriculture, but the breed has fallen into decline in recent decades. With the dairy industry focused exclusively on the quantity of milk produced under confinement conditions, and pricing favoring the quantity of fluid milk over the quality of the components (such as protein and butterfat), the Guernsey's strengths lost market value. Between 1970 and 1990, annual registrations of purebred calves dropped by 60%. Some breeders have sought to stop this decline by increasing milk production, often through the introduction of Red Holstein. This strategy has not been successful at increasing production enough to make the Guernsey competitive with the Holstein, and instead it has caused the breed to lose some of its genetic distinctiveness.

The Guernsey is strong numerically, especially in Britain, but the breed has an uncertain future. It is not clear if the Guernsey can take advantage of the movement to grass-based dairying (as has the Jersey), nor is it known if changes in component pricing may return some of the market value to its milk. Traditionally the breed was a good grazer, able to produce on quality pastures, and adaptable to a variety of climates and conditions. These abilities will be called upon if the Guernsey is to regain its historic place as a leading dairy breed.

Highland

Origin: Scotland
Classification: Standardized breed
Distribution: International
Status: Watch

The shaggy haired, long horned Highland is closely associated with the beauty, mystery, and romance of the Scottish Highlands. Underneath this dramatic appearance lies a useful and productive cattle breed.

The Highland descends from the native cattle of Scotland and is named for the Highland region. The breed was shaped primarily by natural selection, and as a result it is best known for its survival qualities – hardiness, maternal abilities, reproductive efficiency, and longevity. Highland cattle thrive on rough forage and in cold, wet climates. Like the other Scottish beef breeds, the Galloway, Belted Galloway, and Angus, the Highland is celebrated for the excellence of its beef.

The early history of the Highland is not well recorded, though the breed was improved and standardized during the 1800s. Improvement was made through selection alone; the Highland never had any introductions from other breeds. Cattle were raised in the Highlands and on the islands nearby. They were sometimes called *kyloe* cattle, for they swam across the straits (or *kyloes*) on their way to market on the mainland. The breed became well known in Scotland and England, and a herdbook was established in 1884.

Highland cattle were first imported to North America in the 1880s, and importations have continued throughout the 1900s. The breed has always had a small but loyal following, especially in the northern part of the United States and in Canada. It is only recently, however, that Highlands are achieving their greatest popularity. One of the breed's assets is its foraging ability. Highlands consume a wide variety of pest plants as

HIGHLAND CATTLE

well as grass and can be used to improve pastures. The breed is considered a "light grazer" in Europe, used to manage and diversify marginal lands without the negative impact seen with heavier breeds.

Highlands are medium in size, with cows weighing 900–1,300 pounds and bulls 1,500–2,000 pounds. Cattle have long, shaggy coats. Light red is the most common color, but many other solid colors are also seen, including black, brindle, cream, dun, red and white. The horns of the Highland cows sweep out and up dramatically, while those of the bulls are more horizontal with upturned tips.

The Highland is increasing in numbers in North America and globally, with populations found in Australia, Britain, Germany, and elsewhere in Europe. Despite its strength of numbers, however, the Highland faces a threat to its genetic integrity. The European Union (E.U.), an economic organization of most European countries and Great Britain, has issued directives requiring that all livestock herd books in member countries become "compatible." This means that animals recognized by a herdbook in one country must be accorded recognition by herdbooks in other member countries. Many European herdbooks allow upgrading and will register animals with as little as three-fourths pure breeding as purebreds. If E.U. directives force the British studbook to become open to crossbred Highland cattle from Europe, then the consistency and purity of the breed will be reduced. The United States and Canada, whose closed herdbooks remain outside the jurisdiction of the E.U., may soon be the best reservoir of purebred Highland genetics in the world. It is for this reason that the breed, though relatively numerous, has remained a conservation priority, with emphasis placed on the estimated 10,000 Highlands in North America.

Kerry

Origin: Ireland
Classification: Standardized breed
Distribution: Britain, Canada, Ireland, United States
Status: Critical

The Kerry breed was developed in County Kerry in southwest Ireland. The Kerry descends from the ancient Celtic cattle native to the region. Its hardiness and productivity on meager pasture made it the "poor man's cow," and even today the breed is treasured in Kerry.

Kerry cattle were selected for dairy production. The quantity of milk they produce is not as impressive as that of other dairy cattle breeds; what's impressive is that Kerry cows are productive under the most trying conditions and with the lowest inputs of any European dairy breed. Under improved management, production increases.

Kerries are solid black with a small amount of white allowed on the underline. They are small and lean, with cows averaging about 800 pounds and bulls 1,000. The cattle are fine boned with delicate heads and upswept, lyre-shaped white horns with black tips. They are active grazers and browsers, hardy and long-lived.

The Kerry breed was developed in isolation for much of its history. It did not become known in other parts of Ireland and in England until the mid-1800s. A herd book was organized in 1887. An active trade in Kerry cattle, especially in bulls, led to the decline of the quality of the breed in its home region. As a result of this practice and because of the value of the breed, restrictions were put in place to forbid the use of any but purebred Kerry bulls in the county.

Kerry cattle were first imported to North America in 1818, and the American Kerry and Dexter Club was founded in 1911. The breed never became widely popular, however, and its registry closed in 1921. Kerries were reimported to North America in the 1970s. Today the population numbers about 100. The Kerry breed is globally rare, with fewer than 500 cattle alive today, most of them found in Ireland and England.

The Kerry was the genetic foundation of the Dexter, a smaller, dual-purpose breed also developed in Ireland. The Kerry and Dexter

Kerry Cow

MILKING DEVON COW

breeds are closely related, and they were at one time registered in the same herdbook. Bloodtyping has recently established, however, that the two breeds are genetically distinct. They should not be crossbred with each other.

Milking Devon

Origin: England
Classification: Standardized breed
Distribution: United States
Status: Critical

Devon cattle come from the southwestern peninsula of England, where the breed was developed over several centuries. The name comes from Devonshire, though similar cattle were also raised in the neighboring counties of Cornwall, Somerset, and Dorset. Devons were valued for the production of both high quality beef and the rich milk used in Devonshire cream. They were also regarded as the quickest and most active oxen in the British Isles, reputed to trot at six miles per hour while pulling unladen wagons. The breed had the reputation as an easy keeper, able to thrive on rough forage.

The Pilgrims brought Devon cattle with them to New England beginning in 1623. The hardiness and practicality of the breed, combined with the ready availability of Devon cattle near the ports of departure, made Devons an obvious choice for immigrants to the Americas. The breed became well established in New England during the 1600s and spread down the coast as far as Florida during the 1700s and 1800s. Cattle also went west, as Devon oxen were the draft animals of choice on the Oregon Trail. Herdbooks for the breed have been published since 1855.

By the late 1800s, the Devon had gradually been replaced by the Shorthorn, a more productive multi-purpose breed. By 1900, Devon were rarely seen outside of New England. It was only in this region that the breed remained popular, continuing to be valued for hardiness and the ability to thrive under rugged conditions, qualities in which it was superior to the Shorthorn.

By the 1950s, the market for dual-purpose cattle had disappeared, and the Devon breed was nearly extinct. In response to this challenge, the breeder community split. The majority of breeders began to select their animals intensively for beef so that they could compete with other beef breeds. This population was called the Devon or the Beef Devon breed. The remaining breeders continued to select their animals for the traditional purposes: milk, meat, and draft. This population was called the American Milking Devon cattle breed. The American Milking Devon breed is now distinct from other Devon populations in the world and closest to the breed's original type. It is unique to the United States.

The Milking Devon breed barely survived the next two decades, reaching its low point in the 1970s, when fewer than 100 cattle remained. The persistence of a few New England dairy farmers and teamsters protected the breed from extinction. Today, there are over 400 Milking Devon cattle in existence and the population is increasing. The breed is a favorite exhibition animal at historic sites because of its attractiveness and well documented history.

Milking Devon cattle are ruby red with black-tipped white horns. Cattle are medium in size, with cows averaging 1,100 pounds and bulls 1,600 pounds. The appearance is compact yet fine, with a straight topline, square set legs, and well-formed udders. As with most dual-purpose breeds, the Devon population exhibits a range of dairy and beef characteristics. Relatively few strains are being actively selected for dairy production, though this type persists because the lighter dairy-type animals are said to make the best oxen.

Milking Shorthorn

Origin: England
Classification: Standardized breed
Distribution: International
Status: Watch

The Shorthorn is likely the most famous and influential breed of cattle in the history of agriculture. It was among the first livestock breeds to be improved, during the 1700s, and had one of the first herdbooks established, in 1822. Since the early 1800s and until recently, Shorthorns were the most popular cattle in Britain, and they were exported around the world. The story of the Shorthorn has been recorded in countless publications, and images of red, white, and roan Shorthorns dominate sporting art. Yet the breed is now in decline, and its rise and fall is reflective of great changes in agriculture over the past two centuries.

The Shorthorn was historically called the Durham because it originated in the county of Durham. Imported Dutch cattle were crossed with native stocks and selected for performance in both meat and milk production. The breed became

Milking Shorthorn Cow

known early in the 1800s, especially through a travelling promotional exhibit of the famous "Durham Ox." The Ox, calved in 1796, weighed over 3,500 pounds at ten years old and brought much acclaim to the breed.

Shorthorns were first imported to America in the late 1700s, with the largest number of cattle brought in after 1820. The breed was initially concentrated in Ohio and Kentucky, a region rich in grass and corn to feed cattle, but by the end of the 1800s it was spread throughout America. The Shorthorn was valued for its dairy and beef qualities and also used as a draft animal.

Dual-purpose selection has been the major theme of the breed's history. Breeders have always differed in their view of the Shorthorn's purpose, with some selecting for maximum dairy production, others favoring beef, and still others selecting for a balance of the two. At different times, one type of selection would outweigh the other, and then the tide would swing back. For example, the Shorthorn first gained value in the United States as the premiere dairy breed of the mid-1800s, though by 1900 the beef type "Scotch Shorthorns" were the most desirable.

Early in the 1900s, the breed was formally split into a beef type, called the Beef Shorthorn or simply Shorthorn, and a dairy type, called the Milking Shorthorn. Most breeders favored selection for beef, and this trend has continued, especially with the rise of the Holstein as the dominant dairy breed. The Milking Shorthorn, despite its many fine qualities and history of dairy selection, could not compete with the quantity of milk produced by the Holstein, and the breed lost favor.

Another factor in the decline of the Milking Shorthorn has occurred more recently. In an effort to increase milk production, the breed's herdbook has been opened to substantial introduction of outside blood, first from Illawara (Australian Shorthorn) and then Red and White Holstein. Today, many of the bulls registered as Milking Shorthorn are actually one-half or more Holstein. While these introductions did increase production, they also reduced the breed's genetic uniqueness. The breed became less distinct from the Holstein and therefore less useful for commercial crossing. It also lost consistency of performance in its historic traits, such as the ability to produce on grass. This is a major obstacle in its promotion for low-input dairying.

The Milking Shorthorn is medium to large in size, with cows weighing 1,200–1,400 pounds and bulls about one ton. Milking Shorthorns are red, white, roan, or a mixture of the three, sometimes with extensive speckling. Most cattle are horned, though polled strains have been developed.

Although there are several thousand Milking Shorthorns (called Dairy Shorthorns) found in Britain, the breed is declining globally and in North America. Pure American strains are the conservation priority in the United States. These strains will perform well for grass-based dairying, as they are forage-efficient, healthy, long-lived, and productive, with the additional value of high quality beef.

Pineywoods

Origin: United States (Southeast)
Classification: Landrace
Distribution: United States (Southeast)
Status: Critical

The Pineywoods is one of the oldest breeds of cattle in the United States, descending from Spanish cattle brought to the New World beginning in the early 1500s. As the Spanish colonized the Southeast, they established low-input, extensive cattle-ranging systems typical of Spanish ranching. The Pineywoods, Florida Cracker, Texas Longhorn, and other breeds which were developed are called *criollo* cattle, which means "of European origin but born in the New World."

The Pineywoods breed was shaped primarily by agricultural and environmental conditions in Alabama, Mississippi, Georgia, and other parts of the southeastern United States. This, along with some human selection, has resulted in a breed that is heat tolerant, long-lived, resistant to parasites and diseases, and able to be productive on marginal forage. Pineywoods cattle were used in the production of beef, and occasionally used as dairy cattle. They also played an important role as oxen in the timber industry of the Southeast.

Pineywoods Cow

The Pineywoods breed became rare during the late 1800s and early 1900s, when improved English and European cattle were brought into the Southeast. Wherever these breeds could tolerate the environmental conditions, they replaced the native Pineywoods. Improved cattle were also crossed with Pineywoods cows to produce calves which had the survival qualities of the dam and the production qualities of the sire. The overuse of crossbreeding almost eliminated the purebred Pineywoods stocks, and the breed nearly became extinct early in this century.

Only a few families continued to keep purebred Pineywoods herds. As time passed, these herds became more and more isolated from one another, until each herd became a unique and self-contained strain. Today these strains are the most important genetic elements of the breed. The strains are named after the families who conserved them: Holt in Georgia, Barnes in Alabama, and Conway, Bayliss, and Carter in Mississippi. Some of these strains have long and interesting histories. The Carter strain, for example, began in the 1860s when Print Carter, a sixteen-year-old Civil War veteran, swam a herd of red cows over the Pearl River and began raising cattle. The last outside animal to enter the herd was a bull in 1895, yet the herd thrives today.

As one would expect of heat-tolerant cattle, Pineywoods are small and rugged. They exhibit the angular appearance of other Spanish cattle which have become adapted to harsh conditions. Size and selection for beef conformation vary, increasing in the western and northern portions of the range. Cows weigh 600–800 pounds, and bulls average 800–1,200 pounds. A very few "Guinea," or dwarf cattle, are found in the breed. Most Pineywoods cattle are horned, and the oxen have especially impressive horns. Horns vary from long and twisted to short and crumpled.

The Pineywoods breed includes almost all of the solid colors and many of the spotting patterns known to cattle. This is a legacy from Spain, where even today ranchers prefer multicolored herds for range conditions. Some of the family strains have been selected for specific colors or patterns. Conway cattle, for example, are red and white in various patterns. Holt cattle are nearly always black color-sided roan, although James Holt's father preferred solid-colored duns and had entire herds of that color. Griffin cattle tend to be yellow. Other strains, such as Carter and Barnes, include several colors.

The Pineywoods is critically rare, with only a few hundred cattle alive today. Though little known outside the Southeast, the Pineywoods, and the closely related Florida Cracker breed, are high conservation priorities. These American breeds have important qualities, such as fertility and longevity, that are lacking in the Brahman, Zebu, and other heat-tolerant cattle commonly used. The Pineywoods, with almost 500 years of adaptation to the American Southeast, has history on its side as well.

Randall Lineback

Origin: United States
Classification: Landrace
Distribution: United States
Status: Critical

Randall Linebacks are a purebred remnant of lineback-patterned cattle once common in New England. Though the origins of the breed are not clear, it is likely to have originated in New England from a combination of Dutch, English, and French cattle. Historically, Linebacks were multi-purpose, used for dairy, beef, and oxen, and served as an integral part of rural New England life for several centuries. The population was not formally organized, except for the brief existence of the Columbian breed association in the early 1900s. Most of the Lineback population was lost this century through crossbreeding with Holsteins.

The "lineback" part of the breed's name describes the characteristic lineback color pattern. The cattle are blue-black with a white line down their backs. The roan coloring on their sides varies from almost black to nearly white, with black noses, eye rings, ears, feet, and teats. Between the extremes are many animals that are blue roan or speckled. All of the variants are stunning, especially against a background of green grass.

The name Randall comes from the Randall family in Vermont, who kept a closed herd of Linebacks for over eighty years. The Randall herd was one of the few herds of Linebacks that was not crossbred with Holsteins. After the death of Everett Randall, however, the herd was dispersed and most of the animals were lost to slaughter. Through a convoluted pathway and with the efforts of American Livestock Breeds Conservancy members, a small part of the herd was saved and has been the foundation for conservation efforts.

Randall Linebacks are medium in size. Some variation in conformation does occur, but

Randall Lineback Cattle

the majority of cows have dairy conformation and well-developed udders. The bulls are large and demonstrate good growth rates. A few steers have been trained as oxen, a task at which they excel. Their unusual color, willingness, and ability make them attractive and capable draft animals.

The Randall Lineback breed is distinct from the American Lineback, which includes any dairy animal with the lineback color pattern. Though the American Lineback registry includes some animals of historic breeding, this population falls short of the genetic definition of a breed.

Randall Lineback cattle are critically rare, with fewer than 150 animals distributed among a few herds. The cattle are being closely managed for increasing numbers and maintaining genetic health. The breed is more secure now than in the recent past, although its survival is still tenuous.

Red Poll

Origin: England
Classification: Standardized breed
Distribution: Australia, Britain, Canada, United States
Status: Rare

The Red Poll is a dual-purpose breed developed in eastern England in the early 1800s. Red Poll cattle resulted from the combination of the Norfolk and Suffolk cattle breeds. The Norfolk was a small, hardy, red and white horned breed known for the high quality of its beef. The Suffolk was a red, yellow, or brindle polled dairy breed. The resulting breed was recognized in 1846. Its original name – "Norfolk and Suffolk Red Polled cattle" – was shortened to Red Poll in about 1880.

Breeders selected Red Poll cattle for beef and dairy qualities, creating a population that was consistent in appearance, dark red in color, and always polled. A herdbook was established in 1874. The Red Poll remained strongest in its home counties of Norfolk and Suffolk and did not become widely distributed in England. It was, however, exported to several other countries beginning in the late 1800s.

A few Norfolk and Suffolk cattle had been brought to America in the early 1700s, but Red Polls did not arrive until the 1880s. Importations continued through the 1930s. The American Red Poll Association was founded in 1883, five years before the English breed society. Red Polls were efficient and competitive dairy cows, especially

Red Poll Cattle

where total profit over feed costs was measured. They were also known for longevity, often producing 10,000 pounds of milk per year even as teenagers. One cow, Florabel (10862) had an eleven year milk total of 108,310.5 pounds in 1933, which was an all-breed record for the time.

Though these dual-purpose qualities were valuable, American breeders have emphasized the beef characteristics of the Red Poll, especially since the 1960s with the Holstein's dominance in the dairy industry. The Red Poll has responded well to this selection. In studies conducted at the Meat Animal Research Center (MARC) in Clay City, Nebraska, the Red Poll led all breeds in 200-day calf weight per cow *bred*. Red Polls accomplish this by superior fertility, calf vigor, and calf growth rates. Calves are small at birth but grow quickly on their mother's milk. The breed has also been praised for the quality of its beef.

How could a breed with these fine qualities become endangered? The Red Poll is declining in Britain, North America, and globally. Large numbers of Red Poll cows are being lost to the pure breed through their use in herds of commercial cattle and in some other breeds, such as the Salers. It is ironic that while the cattle industry deemphasizes pure breeding, cattle breeders still look to pure breeds to gain the qualities they want in their herds – even if this brings breeds such as the Red Poll to the brink of extinction.

The Red Poll must become more widely recognized if it is to survive. Fortunately, the breed has a great deal to offer. Its long history of pure breeding has given the breed exceptional genetic consistency. Bulls are prepotent, and their offspring are highly predictable and uniform. Distantly related to most other beef breeds, the Red Poll can impart significant hybrid vigor when crossed. The breed was, for example, a foundation of the Senepol breed, developed in this century in the Caribbean.

Crossbreeding serves the breed's interests when bulls are used and can then be acknowledged for the quality of the offspring they produce. Cows, however, must be bred pure to ensure the breed's survival, which means that Red Polls must also maintain their niche as a purebred. This should not be difficult, as the breed is well suited for the production of grass-fed beef.

Red Poll cattle are medium in size, with cows averaging 1,200 pounds and bulls 1,800 pounds. The cattle are solid red, with a little white on the underline tolerated but not desired. They are adaptable and long- lived. Red Polls have quiet dispositions and they are an excellent choice for rotational grazing and other systems where ease of handling is required. Cows can also be used for small-scale and home dairy production. The milk is high in protein and butterfat, making it especially good for cheese. In England, Red Poll cheese is still produced. Wherever these qualities are valued, the Red Poll breed merits attention.

White Park

Origin: England
Classification: Landrace
Distribution: Britain, Canada, United States
Status: Critical

Horned, white cattle are documented in Britain for over 2,000 years, bred and used for ceremonial purposes during the pre-Christian era. In the centuries that followed, herds of feral cattle became scattered throughout England, Scotland and Wales and were hunted as trophy animals. Some of the herds were enclosed into deer parks, such as those at Cadzow, Chartley, Chillingham, Dynevor, and Vaynol, though most cattle were free to roam. As their habitat was gradually reduced, the free roaming cattle disappeared, leaving only the "park" animals as representatives of the ancient type.

These historic cattle attracted attention early in this century. In 1918, the Park Society was formed. It identified nine herds of horned cattle and six herds of polled cattle in its first herd book. British White cattle were originally included, but were reclassified as a separate breed in 1946.

Many of the park cattle were eventually redomesticated, selected for beef production, and crossbred. A few of the herds were conserved as pure stocks. These were the foundation strains for the White Park breed of today. The White Park attracted conservation efforts in the 1970s. When the Rare Breeds Survival Trust was formed as the national livestock conservation organization in Britain, a White Park bull was chosen as its emblem.

The distinct strains of White Parks, as well as the Chillingham (now considered a separate breed), are increasing in number. Bloodtyping has demonstrated the importance of this conservation effort, since remarkable genetic distance exists between White Parks and all other breeds of domestic cattle. This is due to the breed's long isolation and the limited role of human selection in its development.

White Park cattle came to North America during World War II. They were evacuated as part of a British program that sent national treasures to America for safekeeping. The cattle went first to the Bronx Zoo and then to the King Ranch in Texas, where they were maintained as a closed herd for 40 years. The cattle were dispersed in the 1980s, and several conservation herds were established in the United States and in Canada. Breeders formed a North American breed association and a registry to monitor the status of the population, now about 150 cattle. North American breeders have a role to play in conserving some genetic variants that are now rare or have disappeared in the British population.

White Park cattle are athletic and lean in appearance. They are small to medium in size, with cows weighing 800–1,000 pounds and bulls 1,000–1,300 pounds. The light colored, black-tipped horns curve out and then up in a "U" shape. The breed is known for its color pattern, which is called "white park." The coat is white, with black (or red) eye rings, ears, nose, feet, and teats. Many of the animals have dark skin, and some have small speckles of black or blue scattered on the coat. A few individuals are born solid black or red. Historically these were culled, though they are being maintained now as a source of additional diversity to the breed. The white park color pattern also occurs in the British White and in several other breeds of cattle worldwide, as well as in yaks and domesticated gaurs.

The White Park breed should not be confused with the American White Park, a polled beef breed developed in the United States after World War II. This breed originated from the British White and other unrelated breeds. These two populations are quite dissimilar in appearance, history, and use. The White Park breed is sometimes called the Ancient White Park to emphasize its distinctiveness. ❖

White Park Cattle

SAN CLEMENTE GOATS

Chapter 4
Goats

ADAPTABLE, CURIOUS, AND SOCIABLE, goats were among the first animals to be domesticated. Goats' hardiness makes them a part of subsistence agriculture almost everywhere, yet they are also found in highly developed production systems and as pampered companion animals. The global distribution of goats, as well as their many uses, has led to the development of a large number of standardized breeds and landraces, though most of these are not well documented.

Goats are among the most versatile of all domestic animals, used for milk, meat, and fiber, and for packing and land management. Goat milk is inexpensive to produce, easy to digest, and nutritionally rich. Demand for goat milk has encouraged development of many specialized dairy breeds, especially in Europe and India.

Goat meat is consumed worldwide. All goats may ultimately be used for meat, including those which have been culled from dairy and fiber-producing herds. Despite the importance of meat, however, few specialized meat breeds have been developed. Foremost among these is the Boer goat of South Africa, which has recently become established in several other countries, including the United States.

While all goats grow hair coats, some produce specialized and valuable fiber types – such as mohair and cashmere. Mohair is a long lustrous fiber produced by Angora goats. It is similar to the lustrous long wool of sheep, but distinct from the angora fiber produced by Angora rabbits. Mohair is often considered a luxury fiber, although its durability, luster, and strength have made it valuable for a wide variety of uses. Angora goats are named for their region of origin in Turkey, and they are one of the most numerous breeds of goats in the world today. The United States has a large population of Angora goats and is a leader in the production of mohair.

Cashmere is quite different from mohair and is a true luxury fiber. It is the soft undercoat which can be produced by nearly any type of goat. Some goats produce an excellent quantity and quality of cashmere, and while these goats do not constitute a true breed in the genetic sense, they are generally referred to as a breed. Cashmere production originated in Central Asia, but it has now spread to many countries. Though much smaller than the mohair industry, the cashmere industry is expanding. Efforts are underway in Australia and the United States to standardize cashmere production from goats of a variety of breed origins.

The history of goats in North America began with the arrival of Spanish explorers and settlers in the 1500s. Spanish goats provided meat, milk, and leather, which were essential to the agriculture in the regions that became Mexico and the southern United States. English settlers brought a few goats to New England beginning in the 1600s. These two types – Spanish and "Old English" – accounted for most of the goats found in North America until the time of the Civil War. These goats had little market value and were never recognized as distinct breeds.

In the mid-1800s, goat production in the United States began to change. The importation of several goat breeds from Europe increased the breadth of genetic resources available. Angora goats were imported to the United States beginning in the 1850s, and Texas became the center of mohair production. Importation of improved European dairy breeds, including the Toggenburg, Saanen, French Alpine, and Nubian, began about 1900. Dairy production from these breeds far exceeded that of the Spanish and English goats, and these historic types were rapidly crossbred or replaced. The Old English goat became extinct in North America, and it was only in the southeastern and southwestern United States that pure Spanish goats survived.

Recent importations of cashmere goats and Boer goats have also had an impact on goat genetic resources in North America. Imported cashmere goats have been crossed with Spanish goats to increase the production of fiber. Boer goats from South Africa have been imported and crossed with Spanish, Tennessee Fainting, and other breeds to improve meat production. The resulting offspring demonstrate the commercial and biological value of hybrid vigor from the use of two unrelated breeds. At the same time, crossbreeding poses a threat to the survival of Spanish and Tennessee goat breeds, both of which are unique to the United States.

Goat production in the United States has lagged behind that found in other parts of the world, yet demand for the products of this versatile and valuable species is increasing dramatically. The expansion of production and marketing niches for goat cheese, milk, meat, and fiber, as well as the use of goats for pest plant control and land improvement projects show promise for greater appreciation of goat genetic resources. Perhaps this trend can support the conservation of rare goat breeds, especially those which are unique to the United States.

Nigerian Dwarf

Origin: United States
Classification: Standardized breed
Distribution: United States, with a small population in Canada
Status: Rare

Two miniature goat breeds are found in the United States, the Nigerian Dwarf and the Pygmy. These breeds share a common genetic origin in the variable population of small African goats imported to the United States between the 1930s and 1950s. Used originally as exhibition animals in zoos, the goats later became popular as companion animals.

The Pygmy goat breed was recognized by the American Goat Society in 1976. During the last two decades, it has been standardized through selection for small size and stocky conformation. The breed includes a limited color range, primarily agouti, though other solid colors along

Nigerian Dwarf Goat

with the belted color pattern are also accepted.

The Nigerian Dwarf breed originated from the same genetic foundation as the Pygmy, but these goats have been selected to resemble miniature dairy goats with more slender bodies. All colors and patterns, including bi- and tri-color combinations, are accepted. Horns may or may not be present. Nigerian Dwarf goats vary in size, with bucks' height up to 23½" at the withers and does' height up to 22¼". The ideal size of the breed, however, remains a point of debate within the breeder community, and there are now several breed organizations reflecting differing philosophies.

The Nigerian Dwarf was originally selected as a companion and show animal, with emphasis on the breed's graceful appearance and gentle disposition. The production qualities of the breed, however, have also attracted attention. Nigerian Dwarf does produce one to two quarts of milk per day. The milk is high in butterfat and makes excellent cheese and butter. Does generally breed year-round and produce twins. They can be milked for up to ten months, but can also be allowed to dry up on their own if milking is no longer desired. These production qualities make Nigerian Dwarf goats good candidates for small scale milk production where a year-round supply of a moderate amount of milk is the goal.

The versatility of the Nigerian Dwarf, as well as its hardiness and gentle disposition, have given it great appeal, and the breed's population has increased to about 4,000. The breeder community faces a challenge, however, in determining the parameters of the breed. Selection for production qualities may tend to increase the size of the goats, while selection as a companion animal may emphasize small size. Breed conservation will be best served by building consensus around a vision for the breed which includes its unique combination of characteristics.

Oberhasli Goats

Oberhasli

Origin: Switzerland
Classification: Standardized breed
Distribution: International, with the largest population in Switzerland
Status: Watch

The Oberhasli (also known as the Oberhasli-Brienzer) is a dairy breed developed in the mountainous cantons of Bern, Freiburg, Glarus, and Graubunden in Switzerland.

Oberhasli goats were first imported to the United States early in this century, though it was not until 1936 that purebred herds were established and maintained. The breed was initially called the Swiss Alpine. Its registrations were included in the Alpine studbook, and its genetics contributed to the Alpine breed. In 1977, the breed name Oberhasli was adopted, and registration records were separated from the Alpines. This evolution of the breed's name and identification has been one reason that its population in the United States has remained fairly small.

The Oberhasli is alert in appearance with a friendly, gentle disposition. Mature goats are medium in size. Bucks range in height from 30–34", and does 28–32", with weights of 100–150 pounds. While the does are a dependable source of milk, bucks and wethers are also useful as pack animals because of their strength and calm demeanor. Some goat packers prefer Oberhaslis because they are said to be less fearful of water and other trail obstacles than are other breeds.

The breed's color pattern is called *chamoisee*. Goats are brown, with hues between light tan and deep reddish brown, and have black points. Two black stripes from the eyes to the black muzzle give a distinctive facial appearance. The forehead is nearly all black, and black stripes run from the base of each ear to a point just behind the poll and continue along the neck and back to the tail as a dorsal stripe. The Oberhasli has a black belly and light gray to black udder. The legs are black below the knees and hocks and the ears are black on the inside. The Oberhasli face is straight with no evidence of a Roman nose.

The breed is well known internationally, and it is relatively numerous in Switzerland. It is included as a conservation priority in the United States based on the small and isolated population in this country.

San Clemente

Origin: United States (California)
Classification: Feral
Distribution: United States, with a small population in Canada
Status: Critical

The San Clemente is a Spanish goat breed named for its place of origin, San Clemente Island off the coast of California. The ancestors of today's goats were placed on the island by Spanish explorers beginning as early as the 1500s. The Spanish routinely released goats and swine on islands in the New World intending these introductions to serve as a future source of meat.

San Clemente island lacked predators, and the goats thrived during the next several centuries. Recently, however, scientists began to realize the ecological impact the animals were having on the island's indigenous plants and animals. In the mid-1980s, the United States Navy, owner of the island, decided to exterminate the goat population. The Fund for Animals, a New York-based animal rights organization, took legal action to gain the right to remove goats from the island for resettlement on the mainland. Between 1985 and 1986, over 3,000 goats were trapped and removed from the island for adoption.

The Fund for Animals had most of its experience in animal welfare programs, including spay/neuter efforts for dogs and cats. As a result, the Fund required that anyone adopting San Clemente goats have the animals sterilized so that they would not be bred. The policy would have doomed the San Clemente goat breed – just rescued – to extinction. Intervention by ALBC and resistance by a few breeders meant that a small population of breeding animals was retained and became the foundation for breed survival.

The San Clemente is part of the Spanish goat breed family and shares many of the conformational characteristics found in other Spanish goats. For example, the large horizontal ears are usually held forward next to the head, and the horns are large and twisting. The breed has also been shaped by its long history of natural selection on an island. San Clemente goats are consistently small in size (50–90 pounds), as is typical of island-selected animals of any species. The range of colors in the breed is narrow. Most goats are black on the front and reddish tan on the back. The head is black with distinctive pale stripes. Other colors occur, but rarely. This is likely due to the limited colors of the founder population.

The San Clemente goat, unique to North America, is a high conservation priority. Only a few hundred animals remain, and their wide geographic distribution has been an obstacle to the development of a formal breeders network. More people must become involved with this historic and unusual goat breed if it is to survive.

SAN CLEMENTE GOATS

Spanish

Origin: United States
Classification: Landrace
Distribution: Mexico, United States (Texas)
Status: Watch

Beginning in the 1500s, Spanish goats were brought from Spain to the Caribbean Islands and from there to the areas that would become the United States and Mexico. These foundation stocks were an undifferentiated Mediterranean type of goat that was common in Spain at the time but no longer exists. This adds genetic and historic importance to the Spanish goat breeds which evolved in the New World.

Goats flourished in the Americas. Valued as a ready source of milk, meat, and hides, they were taken everywhere the Spanish went and became an integral part of subsistence production across southern North America, Central America, and South America. The use of goats for meat was also important because it allowed cattle to be reserved for draft power, essential for crop production and transportation.

Spanish goats were the only goats known across the southern United States and in most other parts of the Americas for over 300 years. There were many regional types and strains (both domestic and feral) shaped by natural selection and geographic isolation. This changed with the importation of goat breeds from other countries beginning in the second half of the 1800s. The majority of Spanish goats in the United States were crossed with imported breeds for dairy production or were replaced by Angora goats for the production of fiber.

Today, there are relatively few purebred Spanish goats in the United States. One obstacle to the use and promotion of these goats is that the term "Spanish goats" is also used to describe crossbred and nondescript goats of the Southwest. In the Southeast, Spanish goats are also called brush, woods, and scrub goats, and these terms may include both purebred and crossbred animals.

In addition to pure Spanish goats on the U.S. mainland, feral populations of Spanish goats occur on some islands, including San Clemente

Spanish Goat

Island in California, and Mona and Desecheo islands near Puerto Rico. Most of these populations are not well documented, and their relationships to each other are not known. Further research may suggest that all of these strains be grouped into a single breed or that they should be considered several small, closely related breeds.

Conservation of Spanish goats has also been hindered because the breed has no registry or association. Many people still do not understand that the Spanish goat is a pure breed; this is an easy mistake to make given the variable use of the term "Spanish" and the physical variability within the population. Fieldwork in the United States has been helpful in describing the various strains of Spanish goats and in encouraging the development of a formal network of breeders.

Interest in Spanish goats is increasing with the growth of the goat meat industry and with increased recognition of the breed's production qualities. Spanish goats are hardy and rugged, thriving on rough forage and in difficult environments. The does are long lived and prolific. The breed has secured a production niche in Texas, and the largest number of Spanish goat herds are found in this state.

Even in Texas, however, it has become common for Spanish goats to be crossbred for meat production, especially with the Boer goat, a meat breed from South Africa. This cross shows superb hybrid vigor, but the overuse of crossbreeding could threaten the survival of pure Spanish populations. Pure Spanish goats are also being crossbred for the production of cashmere, another useful application of their characteristics but one which also removes animals from the purebred breeding group.

The Spanish goat is a landrace and varies in appearance. The goats range in weight from 50–200 pounds, with the largest animals representing strains that have been selected over many decades for meat production. Spanish goats are usually horned, and the horns on bucks may be large and twisted. The ears are large, and they are held horizontally and forward next to the head. Though the ears may be long like those of Nubians, the ear carriage is distinct, and the straight or concave face is quite different from that of the Nubian breed. Spanish goats can be any color that is found in goats, and color variation, even in single herds, can be dazzling.

The Spanish goat is a global genetic resource that is unique to the United States. The breed's estimated population may be as large as 8,000 goats, but the challenges to its survival are imposing. For these reasons, the Spanish goat is a high conservation priority in the United States.

Tennessee Fainting

Origin: United States (Tennessee)
Classification: Landrace
Distribution: United States
Status: Rare

The goats of this breed have a host of names: Myotonic, Tennessee Fainting, Tennessee Meat, Texas Wooden Leg, Stiff, Nervous, and Scare goats. The names refer to a breed characteristic known as *myotonia congenita*, a condition in which the muscle cells experience prolonged contraction when the goat is startled. The transitory stiffness associated with these contractions can cause the goat to fall down. This is not a true faint, but a muscular phenomenon unrelated to the nervous system. The degree of stiffness varies from goat to goat, with some showing a consistently stiff response and others exhibiting stiffness only rarely.

The breed's history can be traced back to the 1880s. An itinerant farm laborer named John Tinsley came to central Tennessee, reputedly from Nova Scotia. Tinsley had with him four unusual, stiff goats. Goats of this type gradually became known across the region. They were less apt to climb fences and escape from pastures than other goats, and their muscular conformation and high reproductive rate were also valued. Farmers began to appreciate them, and the numbers of "stiff," "nervous," or "fainting" goats increased.

During the 1950s, some Tennessee Fainting goats were taken to the hill country of central

Texas. They were further selected for meat qualities, including larger size, and came to be known as "Wooden Leg" goats.

In the late 1980s, both the Tennessee and Texas branches of this breed were rediscovered. The new enthusiasm for the goats diverged into two major endeavors. One group of breeders worked in the historic tradition, emphasizing the meat qualities of the animals and selecting for growth rate, conformation, and reproductive efficiency. The other group selected for extreme stiffness and small size, promoting the breed as a novelty animal.

As a landrace breed, Tennessee Fainting goats were always variable in size. This variability, emphasized by recent selection, has given rise to a population which ranges in weight from 30–250 pounds. Heavily muscled conformation is consistent among the goats. The ears of Tennessee goats are larger and more horizontal than Swiss breed goats, but smaller and less drooping than Nubian or Spanish goats. The facial profile is usually concave. Most goats are horned, and horns vary from large and twisted to small and simple. While most of the goats have short hair, long haired goats are not unusual and some animals produce cashmere.

Tennessee Fainting goats are found in almost all colors known in goats. Kidding season is always exciting, as new color combinations pop up. Since does like to keep their kids hidden for a few days, looking for these multicolored kids can be like hunting Easter eggs. Does are prolific, with an extended breeding season, and some does will bear kids every six months. Most does produce twins or triplets regularly and have plenty of milk to raise them.

The Tennessee Fainting goat breed is gaining attention for its combination of meat traits with reproductive efficiency, and it is increasingly recognized as an important genetic resource in the United States. Goats are being used both as purebreds and for crossing with other breeds, especially the Boer goat, a recent import from South Africa. While crossbreeding can demonstrate the genetic value of the Tennessee Fainting goat, overuse of purebred does for crossing would threaten the survival of this unique and important American goat breed. It is a high conservation priority. ❖

TENNESSEE FAINTING GOATS

Chapter 5

Horses

"A HORSE! A HORSE! MY KINGDOM for a horse!" said King Richard III near the end of Shakespeare's play about the ill-fated monarch. In the midst of a battle to save his throne, even the King himself was completely dependent upon a mount. Such was the historic value of the horse, giving power, speed, and mobility. Domestication of this species changed the course of human civilization.

For thousands of years, horses have been used in a wide array of environments, working closely with people and serving a myriad of human needs. Desire for aesthetics and style as well as utility shaped the development of domestic horses and produced an easily recognized range of breeds.

Spain and England were the primary sources of horses for North America. Horses from Spain were imported to the Americas beginning in the early 1500s, and Spanish horses were the dominant type in America for three centuries. Spanish genetics were integral to the development of North American gaited horse breeds, such as the American Saddlebred, and the stock horse breeds, including the Appaloosa, Paint, and Quarter Horse. Though far surpassed by these modern breeds today, several pure Spanish breeds still exist in the Americas. They include the Spanish Mustang and Spanish Barb of North America and the Paso Fino, Peruvian Paso, and Criollo of Central and South America.

English horses, including riding and driving types, were brought to North America beginning in the 1600s. The most famous and influential of the English imports was the Thoroughbred breed, which was developed in the mid-1700s and was brought to the United States shortly thereafter.

Two breeds with origins in France have also made significant contributions to horse breeding in North America. The Canadian horse, developed in Quebec from French imports of the 1600s, was widely used in the improvement of other stocks and in the development of new breeds. The Percheron, imported in the 1800s, was once the most popular draft horse breed across the continent.

SUFFOLK HORSES

Horses, like cattle, had the economic value to justify the regular importation of purebred animals and to support pure breeding. Horse breeds and types continuously evolved to meet changing economic and social needs. The diversity of market demands, essentially a variety of habitats, acted to maintain a diversity of breeds. It was not until the mid-1900s that mechanization replaced horse power in agriculture and transportation. This eliminated the markets for many distinct breeds and caused them to decline.

Sport and recreation are now the primary areas of growth in the horse industry. As Catherine Tyler writes, "The changes in horse habitat over the decades of the 20th century ... have shifted the relative proportion of types and breeds. During the peak years of the 1920s, driving horses and ponies, draft horses, and mules formed a large percentage of the North American equine population. Today's recreation-dominated horse activities have resulted in a larger proportion of saddle horses relative to the driving/draft types, with the most popular light breeds comprising the major share of the overall horse population." ALBC's *Taking Stock: The North American Livestock Census* reports that approximately 80% of foals registered annually in North America come from one of six breeds: Appaloosa, Arabian, Quarter Horse, Paint, Standardbred, and Thoroughbred.

Selection within breeds has changed as well. Most riding horse breeds, for example, have been selected according to the same criteria, leading to a convergence of many breeds around the Thoroughbred/Quarter Horse type. Crossbreeding has also been used extensively, and it has been accepted as a quick and easy way to change a breed's characteristics. Thus, most breeds have come to look alike, and many are also related to each other by pedigree.

Relatively few distinct, pure breeds of horses remain, and their conservation should be a high priority for horse fanciers and the horse industry. Rare breeds of horses protect our genetic options for the future as well as the adaptability and utility of this valuable species.

Akhal-Teke

Origin: Turkmenistan
Classification: Landrace in Central Asia, Standardized breed elsewhere
Distribution: Germany, Iran, Russia, Turkmenistan, United States
Status: Critical

Akhal-Teke Horse

The Akhal-Teke descends from the racing horses of Turkmenistan first described over 3,000 years ago. The modern form of the Akhal-Teke dates to about the eighth century, and thus the breed represents the purest living descendant of the ancient Scythian horse. The name "Akhal" means pure, and "Teke" is the name of one of the tribes which has bred the horse for thousands of years.

As Bonnie Hendricks notes in the *International Encyclopedia of the Horse*, Turkmenian horses were historically referred to as "Barb," "Oriental," or "Arab" horses by Europeans, though they were not directly related to the modern Arabian breed. In fact, the Akhal-Teke almost certainly predates the development of the Arabian. Turkmenian stocks had profound influence on the development of horse breeds in Europe, such as the English Thoroughbred. It is now hypothesized that at least two of the three founding sires of the Thoroughbred had Turkmenian blood, either in addition to, or instead of, the previously assumed Arabian background.

For many centuries, the Akhal-Teke remained relatively unchanged. It was found primarily in Central Asia, but also existed as far north and west as Russia, where it was kept in the royal stables. Perhaps the most famous Akhal-Teke was the Russian horse Absent, who won the Olympic gold medal in dressage in 1960 and had a career total of six medals, the most ever earned by any individual horse.

The Akhal-Teke, with its slim build and flowing stride, gives the impression of being an equine version of a Greyhound. Akhal-Teke horses are long in every dimension, with tapering heads, thin necks, and lean bodies. They have strong legs, with dense bone and hard hooves. Smooth and elastic gaits make them appear to float above the ground. The mane and forelock are sparse, the skin is thin, and the coat is short and silky. In winter, a dense undercoat provides warmth. The color gold predominates, with a distinctive metallic sheen characteristic of the breed. The Akhal-Teke averages 15.2 hands (62") at the withers and weighs 900–1,000 pounds. Akhal-Tekes are known as very intelligent horses that respond well to sensitive training.

The Akhal-Teke breed is globally rare, with an estimated population of 1,500–2,000. The largest number of Akhal-Tekes are found in Turkmenistan and in Russia, with smaller populations in Iran, Germany, and the United States. The current U.S. population descends from a few recent imports and numbers about 150 horses.

Akhal-Teke horses excel in many pursuits, including endurance, dressage, and jumping. Despite these abilities and its ancient lineage, the breed has been difficult to promote in North America. Its strange name, unusual body shape and color, and somewhat smaller size mean that the Akhal-Teke does not match the Thoroughbred/warmblood sport horse profile that is popular today. Gradually, however, the success of a few horses is giving the breed more recognition, and a broader base of promotional activity has been established. If successfully conserved, the Akhal-Teke population in North America can be an important genetic reservoir for this ancient breed.

American Cream

Origin: United States (Iowa)
Classification: Standardized breed
Distribution: United States
Status: Critical

The American Cream Draft is the only draft horse breed developed in the United States. American Creams are medium in size, averaging 16 hands (64") at the withers. They are cream in color with pink skin, amber eyes, white manes and tails, and occasional white markings.

The American Cream originated in Iowa early this century, and it has always been rare. The story of the breed begins with a horse named Old Granny, auctioned at a farm sale in Story county, Iowa, in 1911. Old Granny was a cream colored draft mare of unknown ancestry, born sometime between 1890 and 1905. She consistently produced cream offspring. A few Iowa breeders became interested in the cream blood-

AMERICAN CREAM HORSES

line, especially after the birth of the stallion Silver Lace, a great-great grandson of Old Granny, in 1932. Silver Lace was an impressive figure, standing 16 hands high and weighing 2,200 pounds.

During the 1930s, cream draft horses became popular in the counties surrounding Melbourne, Iowa. Clarence T. Rierson, one of the new owners, became interested in the strain and bought all of the mares sired by Silver Lace he could find. He researched the ancestry of each cream horse and recorded their pedigrees. Rierson was one of the founders of the American Cream Draft Horse Association, which was chartered in 1944 with twenty members and 75 foundation horses in the registry. By the time of Rierson's death in 1957, 41 association members had registered almost 200 horses.

Just as the American Cream breed was becoming established, however, the market for draft horses collapsed. Mechanization of agriculture meant that the majority of work horses went to slaughter. The breeding of draft animals nearly ceased. For fourteen years the American Cream Horse Association was inactive, except for the transfer of a single horse. Fortunately, a few people held onto their Creams and thus maintained a slender genetic base which was the foundation for the breed's survival.

In 1982, the Association was reorganized. Breeders worked with the University of Kentucky Equine Blood Typing Lab to determine the breed's genetic parameters. Research results suggested that American Creams were a distinct population within the group of draft horse breeds. This gave great encouragement to the breeders and has played an important part in the breed's revival.

The American Cream Draft Horse Association celebrated its 50th anniversary in 1994. The breed is still critically rare, with fewer than 100 horses alive today, but its numbers are increasing. Promotional efforts are beginning to address the primary obstacle to conservation, the fact that the American Cream breed is largely unknown.

Canadian

Origin: Canada (Quebec)
Classification: Standardized breed
Distribution: Canada, with a small population in the United States
Status: Critical

The Canadian was developed from horses sent from France to Quebec between 1665 and 1670. The limited number of horses in the colony meant that they were highly valued, and geographical isolation meant that the stock remained pure. Over the next century, natural and human selection molded a rugged and versatile breed: strong, dashing, and quick. The Canadian was used for farm work, transport, riding, and racing, earning the nickname "little iron horse." Canadian horses could excel at any task they were put to.

By 1800, the breed had become well known in the United States. Canadian horses were used on stagecoach routes throughout New England. They were also used to improve other stocks and develop new breeds, playing a pivotal role in the formation of the Morgan, the American Saddlebred, and the Standardbred. Though the Canadian horse was celebrated for what it could do, no breed association was ever established in the United States, and there was no documented effort to maintain a purebred population in this country.

The market for Canadian horses was strong throughout the 1800s, and thousands of horses were exported from Canada to the United States. Many of them became cavalry horses and were killed in the Civil War. So many horses were exported or lost that by 1880 the breed was nearly extinct in Canada. In 1886, a small group of breeders in Quebec formed a studbook for the Canadian horse. An association, the Societe des Eleveurs des Chevaux Canadiens, was formed in 1895. Both the federal government of Canada and the provincial government of Quebec maintained breeding programs periodically, but in 1976, fewer than 400 horses remained. Since then, the breed society has been revitalized and has been successful in promoting the breed as a family horse and national treasure all across Canada. There are an estimated 2,000 Canadian horses

Canadian Horse

alive today, and the future of the breed is looking brighter than at any time in the past century.

The Canadian stands 14–16 hands (56-64") at the withers and weighs 900–1,000 pounds. Canadian horses are solid and well-muscled, with a well-arched neck set high on a long, sloping shoulder. The overall impression is one of a round, sturdy, and well-balanced horse. Canadians are primarily black or bay, with full manes and tails. They are energetic without being nervous, and, as appropriate to their heritage, versatile and adaptable for a variety of riding and driving disciplines.

Caspian

Origin: Iran
Classification: Landrace in Iran, Standardized breed elsewhere
Distribution: Britain, with small populations in Australia, Iran, and the United States
Status: Critical

Small, refined horses appear in Persian art and artifacts as early as 3,000 B.C. Among these are the tri-lingual seal of King Darius the Great, the staircase frieze at the Palace of Persopolis, and the Gold Oxus Treasure of Darabgird. The King Darius seal, for example, portrays the King riding a horse so small that his feet almost touch the ground. Until recently, it was believed that such horses had become extinct long ago, but the rediscovery of the Caspian strongly suggests that this breed is a direct descendant of this ancient Persian lineage and one of the primary genetic foundations for modern horse breeds.

The Caspian horse owes its survival to Louise Laylin Firouz, the American-born wife of Narcy Firouz, an Iranian aristocrat. Mr. and Mrs. Firouz went to live in Iran in 1957 and taught horsemanship in Tehran. One of the difficulties they faced was finding small, even-tempered mounts for children. This led Mrs. Firouz to explore rumors of small horses said to be found in the remote villages of the Elburz region above the Caspian Sea. In 1965, she went on an expedition to this region and discovered that the rumors were true. There were indeed a few distinctive and diminutive horses. Mrs. Firouz named the breed the Caspian horse and established a breeding center for its conservation. As the number of Caspians in her herd increased, she was able to spare a few for export, sending 26 horses to Europe between 1971 and 1976. These became the foundation for the breed's survival outside of Iran.

The massive political upheavals of the next 30 years – the overthrow of the Shah, the Islamic Revolution, the Iran-Iraq war, and the resulting economic and social chaos – have created significant obstacles to the Caspian's survival in its native

Caspian Horse

Cleveland Bay Horses

country. The breed has been both honored as a national treasure and disgraced for its association with royalty. Both Mrs. Firouz's first breeding center (which been supported by the Shah) and a second private effort were seized by the radical Islamic government and the horses auctioned. Despite these setbacks, as well as some time spent in prison, Mrs. Firouz continues to live in Iran and work diligently on behalf of the Caspians and other rare breeds which are still found there.

There are about 300 Caspians alive today, with the majority of the population in England. There are about 50 horses in the United States, the results of recent imports. A Caspian registry has been established with the goal of promoting the breed in the North America, and an international Caspian organization has also been formed.

The Caspian has a very distinctive appearance, for though small in size, it is proportioned like a horse rather than a pony. The breed's features include a dished face, vaulted forehead, slim body and legs, high croup, and dense coat. Horses average 11 hands (44") high at the withers and are found in bay, brown, gray, and chestnut. The Caspian breed is being studied in order to document its antiquity, characterize its relationship with other breeds, and describe its place in equine history.

Cleveland Bay

Origin: England
Classification: Standardized breed
Distribution: Britain, with small populations in Australia, Canada, and the United States
Status: Critical

The Cleveland Bay was developed in Yorkshire, England, from a foundation of native horses with some additions of Galloway, Andalusian, and possibly Arabian blood. It was selected for strength, speed, soundness, and stamina, the qualities needed for performance in long distance transport, farm work, and riding over rough terrain. The Cleveland Bay existed in its modern form by 1800, with the breed name "Cleveland" referring to the Vale of Cleveland in Yorkshire, and "Bay" to its color (reddish brown with black points). The Cleveland Bay Horse Society was formed in Britain in 1884.

Two hundred years of pure breeding have resulted in a horse that is genetically consistent. This distinguishes it from the other medium-weight horse breeds known as warm-bloods, which have recently had extensive introductions of Thoroughbred blood. In fact, the breed's genetic distance from the Thoroughbred gives it excellent hybrid vigor in a cross, and Cleveland Bay stallions are often bred to Thoroughbred mares to produce sport horses which combine the best qualities of both breeds.

Cleveland Bays are always bay in color with no white markings except an occasional small star. The horses stand 16–17 hands (64–68") at the withers and weigh 1,200–1,500 pounds. They have well-muscled hindquarters, sloping shoulders, and dense bone. Sound, durable feet are characteristic of the breed, as is a calm disposition. The Cleveland Bay gives an overall impression of dignity and strength.

The breed became rare in this century when mechanization eliminated the demand for farm, coach, and road horses. The growing market for sport horses was of little help, because the lighter and faster Cleveland Bay crossbreds had more value than did the purebreds. This economic reality led breeders to crossbreed their mares. The decline of pure breeding nearly caused the breed to become extinct. By 1960, only six purebred Cleveland Bay stallions remained alive. The British royal family and a few other breeders took the initiative to conserve the breed, and its fortunes have slowly improved. There are perhaps as many as 500 Cleveland Bay horses alive today, with about 50 purebreds in North America.

Cleveland Bay horses were first imported to the United States in the early 1800s. The Cleveland Bay Society of America was founded in 1885, and over 2,000 horses were registered by 1907. Buffalo Bill Cody was among those interested in the breed, and he drove a hitch of four Cleveland Bay stallions in his Wild West Show. The breed's fortunes waxed and waned, however, and the Society became dormant. It was reestablished in 1985 and the numbers of horses and breeders in North America are now increasing.

Clydesdale

Origin: Scotland
Classification: Standardized breed
Distribution: Australia, Britain, Canada, New Zealand, United States
Status: Watch

The Clydesdale is a draft horse breed from Lanarkshire, Scotland, and its name was inspired by the River Clyde. The Clydesdale's history goes back to the mid-1700s, when native stocks were mated with larger Flemish horses and then selected for use in agriculture and for all types of heavy hauling, including transport through the streets of Glasgow. Clydesdale horses were first exhibited under the breed name in 1826, and the breed was further standardized through the system of hiring out purebred stallions. The Clydesdale Horse Society was formed in 1877, and its breed registry records go back to the 1830s.

The Clydesdale was first imported to North America in the 1840s. The breed's attractiveness and size led to its use in promotional hitches, such as one sponsored by the Chicago Stockyard Company. In the 1930s, the Anheuser Busch Company of St. Louis began to use a hitch of Clydesdales to celebrate the end of Prohibition and to represent the company's products. The "Budweiser Clydesdales" were an immediate sensation when they delivered the first post-Prohibition beer to the White House, and the hitch has proven to be one of the most popular and successful advertising symbols of all time. Less well known is the fact that Anheuser Busch has maintained a Clydesdale breeding program since 1953, and that the company's commitment to the breed during the 1950s and 1960s was critical to its survival in North America.

Clydesdale Horse

The Clydesdale combines strength and style. The breed's action is distinctive, with each foot lifted cleanly off the ground so that the bottom of the foot is visible from behind. The forelegs are well planted under the shoulders, the legs are straight, and the feet are durable. Clydesdales should have breadth between the eyes, with a flat facial profile and a wide muzzle. The withers are high, the back short, and the quarters long and well-muscled. The most common color is bay, though black, brown, chestnut, and roan are also seen. The Clydesdale resembles the Shire, and the breeds are related.

As with many breeds of livestock, the Clydesdale has evolved during this century to meet the changing demands of the times. In the early 1900s, the market favored a compact animal. More recently, breeders have selected for a taller horse to be used in fancy hitches and in the show ring. Today's Clydesdales stand 16.2–18 hands (66–72") at the withers and weigh 1,600–1,800 pounds. The largest individuals are taller, weighing as much as 2,200 pounds.

The North American population of Clydesdale horses has increased steadily during the past thirty years, and the population is now at an all-time high, with more purebred foals registered annually than ever before. Globally, the breed is well known but not common, with an estimated global population of about 5,000 horses.

Dartmoor

Origin: England
Classification: Standardized breed
Distribution: International, with most of the population in Britain
Status: Rare

The Dartmoor descends from the native ponies present for many centuries on Dartmoor in southwestern England. Ponies were used for riding, agricultural work, and pulling skips in coal and tin mines. It was customary to range the ponies on the moor when they were not needed, and natural selection helped shape a breed that could thrive on rough terrain and poor forage. This breed is one of five native, regionally-adapted English ponies; the others are the Dales, the Exmoor, the Fell, and the New Forest.

Dartmoor ponies nearly disappeared during the first half of the 1900s. Mechanization eliminated the need for working ponies, and purebreds were crossed for the production of polo ponies and other sport horses. The formation of the Dartmoor Pony Society in the 1930s was a first step in the Dartmoor's survival, although the breed did not begin to increase until after World War II. One of the Society's goals today is the reestablishment of purebred free roaming herds on Dartmoor to replace the mixed breed ponies present now.

Dartmoor ponies average 12 hands (48") at the withers. Bay and brown are the most common colors, but black, chestnut, and gray are also found. White markings are uncommon. The head is small with a broad forehead and small ears. The neck is strong and the quarters are well muscled. The legs have dense, flat bone, and the hooves are durable. Manes and tails are full. The sloping shoulder and long stride give a smooth ride. The ponies' calm, friendly, and willing dispositions make them popular as children's mounts. The breed's consistency of appearance makes it easy to pick Dartmoors out of a crowd and to match ponies for driving teams.

The Dartmoor breed has an estimated global population of 6–8,000. About 200 Dartmoor ponies are found in the United States, descending from imports made during the 1930s, 1940s

DARTMOOR PONY

EXMOOR PONY

and 1950s. The registry of the Dartmoor Pony Society of America was begun in 1936. Demand for Dartmoors exceeds the supply, but the increasing numbers of ponies will gradually make it possible for more people to become involved with this versatile and charming breed.

Exmoor

Origin: England
Distribution: Britain, with small populations in a few other countries
Classification: Standardized breed
Status: Critical

The Exmoor is native to the hill country of Exmoor in southwestern England, a wide expanse of moorland bisected by the River Exe. Exmoors are believed to be the most primitive of the northern European horse breeds, and the breed's antiquity and genetic distance from other breeds has been demonstrated.

Natural selection has shaped the characteristics of the Exmoor. Its winter coat has two layers, an insulating undercoat and an outer coat of greasy, coarse hair that is almost waterproof. This insulation is so effective that snow will collect on the ponies' coats because not enough heat is being lost to melt it; this is called "snow thatching." The body hair grows in patterns that facilitate the shedding of water, and the plentiful mane, tail, and forelock protect delicate parts of the body. The Exmoor has a very large nasal cavity, giving space to warm cold winter air, and a raised rim around the eye protects it against water and debris. The ponies' uniform brown coat is adaptive as camouflage, and no white markings disrupt this concealing color. A lighter brown called "mealy" is found on the muzzle, around the eyes, and on the flanks.

Exmoor ponies average 11.2 hands (46") at the withers and are very stocky for their height. The deep girth and chest make for a large digestive capacity, important for utilizing a large amount of rough forage in the diet. The ponies are athletic, intelligent, and strong willed.

Today, fewer than 800 Exmoor ponies exist in the world, most of them in England. About 150 of them roam free on Exmoor, thriving on their own with no food or shelter provided. Though these ponies live wild, they are all owned, and every fall the herds are rounded up for foals to be collected and sold. These foals are of great genetic value to the domestic population since they retain traits selected for survival in the wild. The Exmoor breed is perfectly adapted to its native environment, so much so that it has been exempted from the general policy of removing all livestock from the moor. In fact, Exmoor ponies are being used as native grazers in several environmentally sensitive areas of England to maintain or increase plant diversity.

The Exmoor is promoted in Great Britain by the Exmoor Pony Society, founded in 1921. The Society maintains information on the status of bloodlines within the breed as well as on its overall numbers. There are about 40 Exmoors in North America, but the numbers are increasing due to recent imports and the work of the Canadian Mountain and Moorland Society, which serves as a breed association for the Exmoor.

Florida Cracker

Origin: United States (Florida)
Classification: Landrace
Distribution: United States (Florida)
Status: Critical

The Florida Cracker horse, like the cattle breed of the same name, traces its ancestry to Spanish stocks brought to the Americas beginning in the 1500s. The formation of the Cracker horse breed was parallel to that of the Spanish Mustang and Spanish Barb further west. The Cracker became a distinctive part of this breed family due to its geographical isolation and location. The long history of horse trading between Florida and Cuba meant a regular and continual introduction of Spanish genetics from Cuba into the Florida population.

Cracker horses were an essential part of the cattle industry in Florida, which began almost 500 years ago and flourishes today. Florida cowboys were nicknamed "crackers" because of the sound made by their whips cracking in the air. This name was also given to the small, agile Spanish horses that they used to work their Spanish cattle.

The Cracker horse suffered a reversal of fortune in the 1930s. The Great Depression led to the creation of a number of relief programs, one of which encouraged the movement of cattle from the Dust Bowl into Florida. With the cattle came the screwworm parasite, and this changed the practice of cattle raising. Before the screwworm, cowboys used horses to herd and drive cattle; with the arrival of the screwworm came the need to rope cattle and hold them for veterinary treatment. As a result, ranchers turned to the larger,

Florida Cracker Horses

Friesian Horse

stronger Quarter Horse, and the Florida Cracker horse declined.

The breed's survival over the last fifty years resulted from the work of a few families who kept distinct blood lines alive in their herds. Several of these lines – such as Ayers, Harvey, Bronson, and Whaley – are still very important within the breed. The Florida Cracker Horse Association was organized in 1989, with its first task the search for remnant herds of Cracker horses. A registry was established, and foundation animals were registered based on their history and external type. Stringent application of the rules has resulted in a very consistent breed. Today, the Florida Cracker is promoted as a valuable and vital part of Florida's heritage, though it is still quite rare with just over 200 horses having been included in the registry.

Florida Crackers are small riding horses, standing 14–15 hands (56–60") at the withers and weighing 800–900 pounds. They have wide foreheads and finely made faces, with straight or slightly roman noses. Croups are sloped and tails are set low. While this external type is distinctive, breed proponents insist that the best way to tell a Florida Cracker horse is to ride one, for its easy, ground-covering gaits rarely occur in other breeds. The breed is found in many colors, with dark bay and black most common, and gray, chestnut, and various shades of dun also present. Enticing leads of paint and roan strains still persist, and these other colors (historically present in the breed) may yet be located in some remote corner of Florida.

Friesian

Origin: The Netherlands
Classification: Standardized breed
Distribution: International, with most of the population in the Netherlands
Status: Watch

The Friesian is a historic driving and riding horse breed from the province of Friesland in The Netherlands. Once seen, the boldness and elegance of the Friesian is unforgettable.

Friesians descend from the "Great Horses" native to northern Europe. These horses were strong and agile enough to carry heavily armored knights on the Crusades. During the 1600s, Dutch horses were crossed with Andalusians from Spain, and the stylish horses that resulted were called Friesians.

Friesians were selected for the action and trotting ability desirable for carriage driving, racing, and circus performance. They were also exported throughout Europe and used to improve a variety of breeds, especially those used in harness. For the past two hundred years, the Friesian breed itself has been kept free from outside blood, making it a genetically distinctive member of the "warmblood" group of horse breeds.

The Friesian nearly became extinct in this century when the market for multi-purpose horses disappeared. By the middle of the 1900s, the population stood at about 500. A riding association called "De Oorsprong" (The Source) was then formed to promote the breed. Members mounted a crusade throughout the country, taking Friesian horses around to remind people of the breed that was their heritage. Interest was rekindled, and today the Friesian is in a period of revival. There are about 8,000 Friesians globally, most of them in the Netherlands.

Friesians are consistent in appearance. The horses average 15–16 hands (60–64") at the withers. They are compact and strong, weighing 1,200–1,400 pounds. The color is solid black, though past registry records also list the color "blue," which may have included blue silver dapple, roan, or possibly gray. The mane and tail are long and wavy, and abundant hair (or "feathering") is found on the lower legs. The head is small and the neck is set high onto the shoulder.

Friesian horses were brought to North America by Dutch settlers in the 1600s. These horses likely had an influence on several American breeds, including the Morgan and the Standardbred. A separate purebred population was not maintained, however, and the Friesian became extinct in this country. Three hundred years later, in the 1970s, the breed was reintroduced to North America by driving enthusiasts. The Friesian Horse Association in the United States was founded in 1983. Only those horses which are approved by Dutch standards may be registered, and stallions to be used for breeding must pass a rigorous inspection of their conformation and performance ability. There are approximately 1,000 Friesians in the United States today, and their numbers are increasing.

Gotland Horse

Gotland

Origin: Sweden
Classification: Standardized breed
Distribution: Denmark and Sweden, with small populations in a few other countries
Status: Watch

The Gotland is native to the island of Gotland in Sweden, where horses of this type have been documented as far back as the Stone Age. Also known as *Russ* or *Skogruss* (meaning "little horse of the woods"), the Gotland is a primitive breed, having been shaped in part by natural selection.

The breed's history is connected with that of the Goths, who inhabited Gotland about 1800 B.C. The Goths used Gotland horses for draft, driving, and transport, and when they migrated into Europe about 200 B.C., Gotland horses pulled their chariots. The Goths eventually travelled into Germany, across the Danube, and into Italy, where they sacked Rome in 410 A.D. The value of Gotland horses to the Goths is recorded in sculpture and in runic inscriptions, but the genetic impact of the breed may also be seen in several European breeds.

Gotland horses remaining on Gotland continued to be used in agriculture and transportation, running loose on the range when they were not needed. During the late 1800s, however, more intensive development of the island led to a shift to larger breeds for heavier work. The market for Gotland horses declined, and the range available to free roaming herds was reduced. By 1900, fewer than 200 horses remained. The Gotland Agricultural Society took an interest in the breed, fencing its range on the island and encouraging pure breeding on the Swedish mainland. The breed has seen a revival in the last few decades and now numbers 6–8,000 globally, with most of the population found in Sweden.

The Gotland is light and elegant in conformation. The breed has pronounced withers, a long back, and a sloping croup. The chest is deep and the shoulders long and sloping. The legs are strong and the feet are strong and hard. Gotlands stand 11.2–13 hands (46–52") at the withers. Dun and bay predominate in the breed, but all colors are allowed except albinos, roans, and piebalds. Characteristic of the breed are a gentle disposition and a lively intelligence. Gotlands are athletic, making excellent trotters and jumpers. Though strong enough to carry adults, they are primarily used for driving and as children's mounts.

Gotlands were first imported into the United States in the 1950s for use in handicapped riding programs. A registry was formed in the 1960s, and over 500 animals have been registered in the past thirty years. The breed almost became extinct in North America during the 1980s. The American Livestock Breeds Conservancy and a small number of breeders have worked to encourage a revival of the Gotland breed. Over one hundred purebred Gotlands are currently found in the United States today, and the population is increasing.

Hackney Horse

Origin: England
Classification: Standardized breed
Distribution: Argentina, Australia, Britain, United States
Status: Rare

The Hackney had its origins among the native horses of eastern England, including the Norfolk Trotter and Yorkshire Roadster from Norfolk, Yorkshire, Lincolnshire, and East Anglia. The introduction of Thoroughbred to this genetic base in the late 1700s gave rise to the Hackney breed.

Roadsters of this time period were expected to have stamina as well as speed. Tom Ryder, who wrote a history of the breed, states that long distance trotting was the horses' strength. For example, Wroot's Pretender, one of the best known of the early Hackneys, once won a match race of seventeen miles by trotting it in less than one hour. Races of 100 miles were not unheard of; Hackneys could cover these distances in less than ten hours.

A breed society to promote the roadster was established in the 1880s, and Hackney was

Hackney Horse

chosen as the breed name. Hackneys were in great demand throughout the 1800s, valued as alert, hardy horses with both speed and style. The coming of the railroad, however, changed the uses of horses significantly. Driving distances became much shorter, and selection for style was emphasized over that for stamina. The result was carriage horses of lighter type and higher action, similar to the pleasure and show types of today. The Hackney retains athletic versatility, however, and the breed can excel in a variety of driving sports. It is also a useful riding horse and often makes a good jumper.

Hackney horses stand 15–16 hands (60–64") at the withers. The body is compact and muscular, with a broad chest and sloping shoulder. The legs are of medium length with large, strong joints, and the feet are durable. The hallmark of the breed is its fluid shoulder action and long, reaching stride. Breed colors are primarily bay, brown, chestnut, and black, with or without white markings.

The Hackney breed is uncommon worldwide, with a global population of perhaps 3,000. The majority of the horses are found in Britain, with fewer than 200 Hackney horses in North America and about 300 in Argentina. The breed type varies from place to place and strain to strain. Most strains have had recent introductions of Thoroughbred blood, and the British studbook allows a limited amount of this crossing. Relatively few strains retain the pure genetics of the historic Hackney breed, and these are a conservation priority.

The Hackney horse is related to but distinct from the Hackney pony, which was established in the 1870s through crossing of Hackney horses with Welsh, Fell, and other pony breeds. The pony breed is especially popular in North America, where it is numerous and not in need of conservation.

Lipizzan

Origin: Austria
Classification: Standardized breed
Distribution: International
Status: Rare

The Lipizzan is one of the most famous horse breeds in the world. Celebrated on film and television, the "white stallions" of the Spanish Riding School are easily recognized by the public.

The Lipizzan breed had its genetic origins in Spain, from stock which included both Spanish and Moroccan Barb blood. Spanish horses were used for all purposes, but their athletic skill and intelligence was especially valued in the classical riding practiced in the Renaissance. For this reason, Maximillian II of Austria (the son of Emperor Ferdinand I) chose Spanish horses as the foundation for a royal stud formed in 1560 at Kladrub. Maximillian's brother Archduke Charles also established a royal stud with Spanish stock about 1580 in the village of Lipizza near Triest. The two studs collaborated closely, and both supplied horses to the Spanish Riding School of Vienna. Eventually selection goals diverged, which led to the formation of the Kladruby breed as a heavier carriage horse and the Lipizzan breed as a lighter riding horse.

The history of the Lipizzan is reflected in the historic documentation of bloodlines within the breed. Six stallion families are known: Pluto, Conversano, Favory, Neapolitano, Siglavy, and Maestoso. There are several mare families as well. These bloodlines are identified genetic resources within the breed, and their conservation is of great importance.

The Lipizzan is a compact and powerful horse standing 14.2–15.2 hands (56–60") at the withers. The head is long with a flat or convex profile, and the neck is muscular and well arched. The chest is deep and wide, with powerful and sloping shoulders. Gray is the predominant color. Most foals are born dark and lighten to white between the ages of four and ten, although a few remain black or bay as adults.

The fortunes of the Lipizzan in this century have been threatened by economic and political instability and war in Central Europe. The Hapsburg empire collapsed in 1910, and Lipizza found itself a part of Italy. Part of the Lipizzan horse population was moved to Austria, where the government established a stud at Piber. World

Lipizzan Horses

Mountain Pleasure Horse

War II caused significant damage to the breed, with the loss of a great many horses and the pedigree records of many of the survivors, which were then removed from the purebred population. More recently, civil war in the former Yugoslavia in the 1980s and 1990s destroyed many herds of that region. Time and again, the Lipizzan's international distribution has proven essential to its survival.

The Lipizzan breed is rare, with an estimated global population of about 3,000 horses, including about 600 in North America. This population originated from imports of the last 30 years. The breed is most often used as it was traditionally, for riding and driving, though it has just begun to gain success in dressage competitions. Though the superior balance and action of the Lipizzan has been celebrated for centuries, the breed gives a very different performance than the larger "warmblood" breeds of Europe. As a result, it is not always rewarded in the showring. If Lipizzans find a modern niche as sport horses, continued selection for performance will be encouraged, and conservation of the breed's complete genetic heritage will be accomplished.

Mountain Pleasure

Origin: United States (Kentucky)
Classification: Landrace
Distribution: United States
Status: Critical

The limestone plateau west of the Appalachian Mountains of the United States has long been known as horse country. Horse breeding has been an important economic enterprise, and horses have been used for riding, agricultural work, and pulling a variety of vehicles. They had to be easy keepers, rugged, surefooted, and willing. Kentucky was also important because of its central location, where Spanish horses from the Southeast and Southwest could easily be crossed with English and European horses from the East.

History suggests that Spanish horses likely contributed the "gaited" characteristic found in many of the breeds developed in this region. Gaited horses are those which naturally have gaits other than (or in addition to) the walk, trot, and canter of all horses; they may include the rack, single foot, and running walk. These gaits are more comfortable for the rider than the trot.

The Mountain Pleasure horse is a landrace which has been relatively unchanged for a century or more. It reflects the primitive Appalachian gaited horse type and may be ancestral to modern breeds developed in the region during the late 1800s and early 1900s, including the American Saddlebred and the Tennessee Walking Horse. As a landrace, the Mountain Pleasure is variable in type, with some horses having distinctively Spanish features and others resembling the larger, modern breeds. Consistent among all is a smooth four-beat gait that replaces the trot. The horses stand 14.2–15.2 hands (58–62") at the withers and weigh 850–950 pounds. Most of the solid colors known in horses occur in the breed, including grays and roans. Spotted individuals are not accepted by the registry.

Mountain Pleasure horses are prized for their calm temperament, and the breed is best suited to be a family pleasure horse rather than a high powered show animal. The most famous Mountain Pleasure horse, however, was Roy Rogers' original Trigger. Since Trigger did not trot, he was always filmed at a walk or a canter.

The Mountain Pleasure Horse Association was formed in 1989 to conserve and promote the breed in North America. Since then, about 1,800 horses have been registered. The Association aims to maintain the heritage of the breed, with emphasis on its kind disposition and comfortable natural gait. The Mountain Pleasure is closely related to the Rocky Mountain horse and many horses are included in registries of both breeds.

Rocky Mountain

Origin: United States (Kentucky)
Classification: Landrace
Distribution: United States
Status: Rare

The Rocky Mountain horse breed actually hails from the Appalachian region of eastern Kentucky, and it was one of the foundations of the family of breeds which developed there. Kentucky was geographically significant, as its central location meant that Spanish horses from the South and Southeast could easily be crossed with English horses from the East. From this genetic combination would come all of the gaited breeds developed in America, including the American Saddlebred and Tennessee Walking Horse.

Gaited horses are those which naturally have gaits other than (or in addition to) the walk, trot, and canter of all horses; they may include the rack, single foot, and running walk, which are more comfortable for the rider than the trot. Spanish horses likely contributed this characteristic to the American breeds.

Rocky Mountain Horse

The Rocky Mountain breed was also shaped by its early uses in the limestone plateau and mountains of eastern Kentucky. Horses were expected to be adaptable and versatile, easy keepers, rugged and surefooted, willing in disposition. The Rocky Mountain breed reflects the primitive gaited horse type and may be ancestral to modern breeds developed later.

The breed name distinguishes a specific strain of gaited horses that descended from a stallion brought into eastern Kentucky from the West around the turn of the century. The stallion was bred to native mares, and the most famous of the offspring was Old Tobe, considered to be the foundation sire of the breed. Old Tobe was owned by Sam Tuttle, who for many years operated the concession for horseback riding in the Natural Bridge State Park in eastern Kentucky. Tuttle put the most timid riders on Tobe, for he had smooth gaits and a gentle temperament. Tobe was also used as a breeding stallion until the age of 37 and figures in the pedigrees of many of today's Rocky Mountain horses.

The Rocky Mountain Horse Association was formed in 1986 through the efforts of Rea Swan of Lexington, Kentucky. She worked for several years to understand breed history and locate remnants of the breed and establish a registry to record pedigrees. The Rocky Mountain breed has grown quickly, as it fits the current market for easy riding, gentle family pleasure horses, and now numbers about 3,000.

As a landrace breed, the Rocky Mountain is somewhat variable in type. Some horses have very Spanish features while others appear more like the larger, modern breeds. Consistent among all is a smooth four-beat gait that replaces the trot, and a calm, friendly disposition. The horses stand 14.2–16 hands (58–64") at the withers and weigh 850–1,000 pounds. Most of the solid colors known in horses occur in the breed, including gray and roan, though chocolate of varying shades with a flaxen mane and tail is the most popular. Spotted individuals occur occasionally, but they are not accepted by the registry.

SHIRE HORSE

Shire

Origin: England
Classification: Standardized breed
Distribution: Australia, Britain, United States
Status: Watch

The Shire horse gets its name from the rural "shires," or counties, in England, where it was traditionally used for agriculture, heavy hauling, and war. The breed descends from the "Great Horse" of medieval England, a large, strong mount which could carried a knight and armor weighing up to 400 pounds with swiftness and agility. These native English horses were crossed with Flemish stallions in the 1200s and with Dutch Friesians in the 1500s. From that period on, the region of Lincolnshire was home to a distinctive large, black horse called the Lincolnshire Black or Heavy Black.

Large horses were used for long distance hauling and heavy transport over rough country roads and uneven cobblestone streets of cities, and the Heavy Black gradually spread across England in the 1700s. Modern Shire type was established about 1790 based on work done by pioneering livestock breeder Robert Bakewell, also known for his work on Leicester Longwool sheep and English Longhorn cattle. Further standardization was undertaken in the late 1800s, with the formation of a breed society in 1878 and the christening of the breed name Shire in 1884.

The Shire peaked in popularity between 1880 and World War I, but its population plummeted with the mechanization of agriculture and transportation after World War II. The breed declined to its lowest point in the early 1960s, but its fortunes have improved considerably since then. Like the Clydesdale, the Shire has benefitted from its use in promotional hitches, especially those advertising breweries. The Whitbread's hitch of gray Shires, for example, is as well known in England as the Budweiser Clydesdale hitch is in the United States.

Shires were first imported to the United States in the mid-1800s, and the American Shire Horse Association was founded in 1885. The registry has operated continuously except for 1955–1961, when most activity ceased and the breed almost became extinct in the United States. The Shire then experienced a slow revival and today it is more numerous than it has been in many decades, both globally and in the United States.

Shires average 17.2 hands (70") at the withers, sometimes reaching 19.2 (76"), with a weight of greater than 2,000 pounds. Bay, black, brown, and gray are the most common colors in the breed. The head has a convex profile with a broad forehead and long ears. The neck is long and well arched, with a long shoulder. The legs are short and feathered below the knees and hocks. The overall impression of the Shire is one of substance and power.

Spanish Mustang/ Spanish Barb

Origin: United States, Mexico
Classification: Landrace, with some Feral populations
Distribution: United States, with a small population in Canada
Status: Critical

The Spanish Mustang and the Spanish Barb are closely related breeds that descend from horses brought by Spanish explorers and colonists to the Americas beginning in the 1500s. Horses were an integral part of Spanish success in the New World, and both domestic and feral stocks spread far and wide. For centuries, Spanish horses were the most common type of horse throughout the Southeast and all of the regions west of the Mississippi. Beginning in the mid- to late 1800s, however, almost all Spanish stocks were crossbred with or replaced by horses of larger size, including Thoroughbreds, other riding horse breeds, and draft horses. Ubiquitous in 1750 and 1850, the pure Spanish horse in North America was almost extinct by 1950.

A few herds of pure Spanish horses

Spanish Mustang/Spanish Barb Horses

remained in the Southeast and the Southwest, owned by traditional ranchers or Native American tribes, and a few were found in isolated feral herds. From these remnants, conservation programs began. Several groups of people have been involved. First among these was the Spanish Mustang Registry, founded in 1957, followed by the American Indian Horse Registry, the Spanish Barb Breeders Association, and the Southwest Spanish Mustang Association. Today the breed population numbers about 1,500–2,000 horses.

The North American Spanish horse population includes many distinct strains, and these could be considered either parts of a larger, single breed, or several smaller, closely related breeds. Given the overall consistency of this population and the fact that many individual horses are registered in more than one of the registries, the American Livestock Breeds Conservancy has chosen to consider them as one breed. The exception is the Florida Cracker horse, which is considered a separate breed because of its geographic isolation and the fact that few if any horses are double registered with the other groups.

Spanish Mustangs have conformational features that distinguish them from other riding breeds. They are small, standing 13.2 to 15 hands (54–60") at the withers and weighing 700–900 pounds. The horses are generally short coupled and deep bodied, but narrow from the front so that the front legs join the chest in the shape of an "A" rather than the shape of a "U" that is seen in the stock horse breeds. The croup is sloping and the tail is set low. The horses have broad foreheads and narrow faces, and the profiles may be either straight or convex. Spanish horses are athletic and useful for a variety of riding disciplines, among them ranch work, endurance competitions, and pleasure riding. They have an unusually long stride, and many of them are gaited.

Nearly all colors of horses occur in this breed. Solid colors include black, grullo, bay, dun, buckskin, chestnut, red dun, palomino, and cream. Gray and roan are common, as are every sort of paint (frame, sabino, and tobiano) and the leopard complex of colors commonly associated with the Appaloosa breed (including blankets, leopards, and varnish roans). Indeed, Spanish genetics are the source of the color in the Paint, Pinto, Appaloosa, Pony of the Americas, Palomino, and other breeds. The Spanish horse has also made substantial contributions to the American gaited breeds and to the American Quarter Horse and other stock horse breeds.

Suffolk

Origin: England
Classification: Standardized breed
Distribution: Britain, United States, with a small population in Australia
Status: Critical

The Suffolk is the only draft horse breed developed and selected exclusively for farm work. The breed originated in Suffolk and Norfolk in eastern England, where horses of draft type date back to the 1500s. The Suffolk Horse Society was founded in 1877, and the first volume of the Suffolk studbook was published in 1880. All horses in the studbook trace their male lines back to the foundation stallion Crisp's Horse of Ufford, foaled in 1768.

Suffolk and Norfolk are bordered on the north, east, and south by the North Sea, and on the west by the fens (or marshes). The region is so isolated, and the Suffolk horse was considered so valuable that it was bred pure and changed little over the past four hundred years. The Suffolk is unusual among draft breeds in that it was never selected for anything other than agricultural work. As a result, the breed retains characteristics relevant to this enterprise, such as the strength and stamina to plow through heavy clay, hardiness, a willing disposition, and easy-keeping qualities.

Suffolk horses were first imported to Canada in 1865 and to the United States in 1880. The Suffolk had strongholds in the Midwest, New England, and Ontario, though it was never as popular as the Percheron, Belgian, or Clydesdale. The breed declined in numbers after World War II and came close to global extinction in the 1950s. It was only through the efforts of a few breeders that the Suffolk survived. Today there are about 600 Suffolk horses in the United States and about 200 in England. The breed is still rare, but its numbers are increasing. The Suffolk's moderate size compared to that of the other draft breeds

Suffolk Horses

has been an asset in its promotion to those who continue to farm with horses.

The Suffolk breed is consistent and distinctive in appearance. All Suffolks are chestnut in color, one of seven shades between gold and liver. White markings are rarely seen. The horses stand 16–17 hands (64–68") at the withers and average 1,800 pounds. The legs are short, clean, and well-muscled, with dense bone. The shoulders are inclined to be upright, suitable more for power than for action. The back is short and strong, and the hindquarters are long and smooth. The breed's nickname "Suffolk Punch" refers to the horses' rounded appearance.

The beauty of the Suffolk is summed up by the writer Marguerite Henry in her classic *Album of Horses*: "His color is bright chestnut – like a tongue of fire against the black field furrows, against green corn blades, against yellow wheat, against blue horizons." Today's Suffolks reflect not only the breed's history but also the long tradition of farming with horses. They are a living treasure. ❖

Chapter 6

Pigs

THE RAISING OF PIGS has crossed almost every geographic and cultural boundary in the world. Highly adaptable, hardy, and prolific, pigs are one of the most globally important and numerous of all domestic animals. Domestication occurred about 10,000 years ago, taking place in several locations and including many different types of pigs. Most of these foundation stocks were varieties of Eurasian wild pig (*Sus scrofa*), but other species were also domesticated.

Pigs were traditionally raised with little direct management. They were allowed to forage for their own food, including acorns and other nuts, roots, fruit, grain, grass, rodents and snakes. They were also fed by-products from dairies, bakeries, and breweries, and used to glean fields after harvest. Pigs' ability to utilize spoiled food and other waste made them a valuable sanitation service even in urban settings. Under such "extensive" husbandry, pigs could be used to produce meat with little investment in feed, time, or labor.

The management of pigs remained essentially unchanged for thousands of years. In the last 200 years, however, pig production has gradually become a highly specialized industry. Today, most pigs in North America are kept in large, climate-controlled buildings and fed high-energy grains that have been grown and transported specifically as animal feed. The evolution of pig husbandry has affected the number and type of pig breeds that were raised.

Pigs were first brought to the Americas by Columbus in 1493, DeSoto in 1539, and other early explorers. As colonies were established, pigs were imported from England, Spain, Portugal, and other countries; for example, 600 pigs were sent from England to Jamestown in 1608. The types of pigs imported were not well documented until after about 1800, when a multitude of breed names began to appear in the historical record to describe both European and American stocks. These include the Bedford, Byfield, Essex, Grass, Jersey Red, Leicester, Red Guinea, Russian, Spanish Black, Suffolk, and Welsh. The most important were three imported breeds – the Berkshire, the Big China, and the Irish Grazier – which were widely used as improvers of common stocks.

TAMWORTH AND RED WATTLE PIGS

Pigs were an essential part of every farm, used for home and community production of lard and pork. They could also be driven to market to generate ready income. Every region seemed to have its own breed of pigs, selected from available stocks to fit the specific climate, uses, and markets. For most of the 1800s, decentralization of breeds and breeding was characteristic of pig production, and this served to maintain a broad genetic foundation for the future.

The roots of centralization, however, were already present. The gradual transition from subsistence to commercial agriculture changed the types and amounts of crops raised. As the larger settled farms of the Midwest began to produce excess corn, the availability and low cost of this feed attracted pig production and processing to the region; it soon became known as the Corn Belt. By the mid-1800s, the states that produced the most corn also produced the most pigs, and production was declining in the East and New England where costs of feed and other production elements were higher.

At the same time that the industry was becoming geographically centralized, the number of breeds of pigs began to decline. Several breeds became extinct by the early 1900s. Some of these were combined to develop new breeds. The Chester White, Jersey-Duroc, and Poland China, for example, were developed between 1840 and 1900 using the broad genetic base provided by a combination of several other breeds; the founding stocks were lost as distinct populations.

Pig breeds were traditionally classified as one of two types, lard or bacon, depending upon their primary use. Lard breeds were used to produce lard, a cooking fat and mechanical lubricant. These pigs were compact and thick, with short legs and deep bodies. They fattened quickly on corn, and their meat had large amounts of fat in it. This was considered desirable for improved taste and keeping qualities of the pork. In contrast, bacon pigs were long, lean, and muscular. They were traditionally fed on legumes, small grains, turnips, and dairy by-products, feeds which are high in protein and low in energy. As a result, bacon pigs grew more slowly and put on more muscle than fat. Almost all American pig breeds were considered lard types, with only the Yorkshire and the Tamworth classified as bacon breeds.

The market for lard was very strong during World War II, when it was used in the manufacture of explosives. With most lard diverted for military purposes, people had to switch to vegetable oils for cooking. After the war, these oils were successfully marketed as healthier fats, and lard never regained its place in the diet. About the same time, petrochemicals and synthetic nitroglycerine replaced lard for industrial and military purposes. With the decline in the market for lard, demand for lard pigs collapsed.

This sudden market shift caused selection of the lard pig breeds to change completely. Breeders needed to produce leaner meat, and they began to select pigs for muscling, rather than fattening, when fed corn. The most popular breeds of the time, including the Berkshire, Duroc, Hampshire, Poland China, and Yorkshire, received most of the industry's attention. These breeds were widespread, and they had the genetic diversity necessary for selection to change direction. The less popular lard breeds were ignored, and most of them disappeared. Only three breeds of traditional lard type remain today, the Choctaw, Guinea Hog, and Mulefoot. These breeds survived in isolated parts of the Southeast, where small farm production continued longer and traditional types retained more value.

The economic structure of pig production has also changed dramatically since World War II. Pigs used to be an integral part of subsistence and diversified commercial farming, and every farmer kept a few pigs. In contrast, production today is dominated by a few large corporations which are vertically integrated, controlling every step from selection of breeding stock to marketing pork in the supermarket. The number of pig farmers has declined precipitously; some farmers have become contract growers for the corporations and others have gone out of business. Fewer and fewer sale barns and processing facilities are open to independent pig farmers.

This economic centralization has led to genetic narrowing as well. Today, the pork industry rests on a three-way cross between a few highly selected strains of the Duroc, Hampshire, and Yorkshire breeds which have been chosen for performance under intensive husbandry. The seven other commercial breeds that remain, including the Berkshire and Poland China, have declined in economic importance, and a handful of critically rare breeds barely cling to survival. It is ironic that an industry whose success was built on a foundation of genetic diversity is rapidly discarding most of its remaining genetic material.

Rare breeds of pigs will not be conserved by the pork industry, since these breeds do not have the characteristics desired for today's intensive production of pork. Instead, survival of rare breeds depends upon their use in traditional production niches, such as on pasture and in woodlands, where their hardiness and efficiency have economic value. The market for sustainably-produced pork is very small, but it is increasing. In addition, diversified commercial farmers are once again becoming interested in using pigs in conjunction with land clearing, field crops, specialty vegetables, dairying, fruit production, and other agricultural enterprises. This may also provide habitat for rare breeds.

Choctaw

Origin: United States
Classification: Landrace
Distribution: United States (Oklahoma)
Status: Critical

The Choctaw is an American breed that descends from Spanish stock brought to the Americas beginning in the 1500s. Spanish pigs were used by Native Americans, European settlers, and a succession of other peoples in the Southeast for over three hundred years.

When the Choctaw and other Native Americans first migrated from the deep South to the Oklahoma Territory in the early 1800s, they took their animals with them, and this was the beginning of the Oklahoma Choctaw pig population. In the 1830s, the United States government removed the Five Civilized Tribes from

Choctaw Pig

Mississippi and Alabama, forcing them to move to Oklahoma; hogs were also taken in this migration. The modern breed's name comes from the Choctaw tribe, with whom it has been associated for so long.

Choctaw hogs have changed little in appearance during the past 150 years. They are small in size, weighing about 120 pounds, and are black with occasional white markings. The ears are erect to slightly drooping. Many Choctaws have fleshy wattles on each side of the neck. The toes are fused, forming a single, mule-like hoof. The combination of the distinctive hoof and wattles are consistent with the breed's Spanish origin, and these traits are often found in populations that descend from Spanish stocks. The Mulefoot hog, for example, shares some of its physical characteristics with the Choctaw, though the two breeds have diverged over the past century and are now distinct from each other. The Choctaw, as a pure Spanish breed, is also distinct from the "Pineywoods Rooter" and other genetically mixed feral pig populations still found in the Southeast.

The Choctaw breed was shaped by natural selection, and it is obvious that Choctaw pigs are built for survival. They have long legs and are heavier in the forequarters than in the rear, making them both fast and athletic. This conformation has its advantages in the wild, but it does not produce much of a market carcass.

Today, Choctaws in Oklahoma are managed as they were traditionally. They are earmarked and then allowed to run free on the open range. The hogs require little care and are able to forage for their own food, including roots, plants, acorns, and berries. Periodically, the hogs are rounded up, usually being trailed with dogs and then trapped. Once captured, the young ones are ear marked and sorted. Those slated for market are confined and fed on corn to increase their weight. Those not needed for sale are released, zipping back into the woods as quickly as they can. People familiar with docile barnyard pigs are amazed at the speed and agility of Choctaw hogs. At the same time, Choctaws kept in confinement can also become quite tame.

The Choctaw breed is critically rare, with a population estimated at a few hundred animals. Almost all of them are found in the few counties of southeast Oklahoma that were formerly the Choctaw Nation. Few commercial opportunities are available for the breed's use, and lacking any

Gloucestershire Old Spots Pigs

economic return to support conservation efforts, the Choctaw faces substantial obstacles to survival. Nevertheless, the breed is a high conservation priority because it has qualities which may be of importance in the future.

Gloucestershire Old Spots

Origin: England
Classification: Standardized breed
Distribution: Britain, United States
Status: Critical

The Gloucestershire Old Spots is a historic pig breed known for its distinctive white coat with black spots. The breed was developed in the Berkley Vale of Gloucestershire, England, during the 1800s. Its origins are not known, though it was likely based on the Cumberland, Lincolnshire Curly Coat, and several other white breeds, most of which are now extinct.

Gloucestershire (pronounced *Glostersheer*) pigs were selected as excellent foragers and grazers. The pigs were thrifty, able to make a living from pasture and agricultural by-products, such as whey from cheese making, windfall apples in orchards, and the residue from pressing cider. These easy-keeping qualities gave Gloucestershire Old Spots the nicknames "cottage pig" and "orchard pig." Apple-flavored pork from Gloucestershire was a traditional seasonal treat in the markets of London.

The Gloucestershire Old Spots breed became rare after World War II, when the shift to intensive pig production reduced interest in pigs which could thrive out of doors. The breed nearly became extinct in the 1960s, though its population has increased slowly since then. Today, there are about 500 Gloucestershire Old Spots in Britain. The breed is primarily used to increase maternal qualities in commercial herds, and its performance value in this system has improved the market for purebred pigs.

Gloucestershire Old Spots pigs were imported to the United States during this century, and they made genetic contributions to several American breeds, especially the American Spot and the Chester White. The breed never became numerous in the United States, however, and was practically extinct by the 1990s. In 1995, Kelmscott Farm Foundation of Lincolnville, Maine, organized an importation of twenty Gloucestershire piglets to reestablish the purebred population in America. A breed society has been founded, and the number of animals is increasing. This conservation program, supported by the American Livestock Breeds Conservancy, has once again made the Gloucestershire Old Spots pig breed available as a genetic resource in North America.

The Gloucestershire Old Spots pig is notable for its docility, intelligence, and prolificacy. The breed's maternal skills make it able to raise large litters of piglets on pasture. Its disposition and self-sufficiency should make it attractive for farmers raising pasture pigs and those who want to add pigs to diversified operations.

Guinea Hog

Origin: United States
Classification: Landrace
Distribution: United States
Status: Critical

The Guinea Hog is a small, black breed of swine that is unique to the United States. Also known as the Pineywoods Guinea and Guinea Forest Hog, the breed has a history that is both mysterious and tantalizing. Though it was popular for a long period of time in America, today the Guinea Hog is nearly extinct.

Pigs were imported from West Africa to America in conjunction with the slave trade. The imports were documented as early as 1804 by Thomas Jefferson and other Virginia farmers. These large, square animals were called Red Guineas, because they had red or sandy colored hair. Red Guineas were common throughout the mid-Atlantic region during the 1800s. The breed disappeared as a distinct population, however, in the 1880s, when most of the red breeds and types

Guinea Hog

of pigs in the eastern United States were combined to form the new Jersey-Duroc breed.

The name Guinea occurs again a few decades later in the southeastern United States, though describing a different animal entirely – a small, black pig common on homesteads across the region. Guinea Hogs were expected to forage for their own food, eat rodents and other small animals, grass, roots, and nuts, and clean out garden beds. The hogs were also kept in the yard where they would eat snakes and thus create a safe zone around the house. These Guineas were hardy and efficient, gaining well on the roughest of forage and producing the hams, bacon, and lard essential for subsistence farming.

Guinea Hogs were widespread, and descriptions of them varied. Generally, the hogs were small, weighing 100-300 pounds, and black or bluish-black in color. They had upright ears, a hairy coat, and a curly tail. Beyond this, conformation varied, as hogs could have short or long noses and be "big boned," "medium boned," or "fine boned." It is likely that many strains of Guinea Hogs existed. Since most of these are extinct, it is now impossible to weave together all the threads of the Guinea Hog story into a single neat piece.

The Guinea Hog became rare in recent decades as the habitat of the homestead hog disappeared, and it survived only in the most isolated parts of the Southeast. During the 1980s, several new herds of Guinea Hogs were established, partly in response to the pet pig market. The Guinea Hog Association was formed in 1990 to focus attention on the breed and promote its survival.

Several mysteries confuse the breed's history. The relationship between the historic Red Guinea and the Guinea Hog may be simply the common use of the term "guinea" to refer to an African origin. "Guinea" may also refer to the small size of the pigs, somewhat akin to the description of miniature Florida Cracker and Pineywoods cattle as "guinea cows." The Guinea Hog may or may not be related to the Essex, a small, black English breed which was imported to the United States in about 1820 and used in the development of the Hampshire. Essex pigs were found in the Southeast about 1900, though the breed's history is obscure and it disappeared this century. "Guinea Essex" pigs were used in research at Texas A & M University and at the Hormel Institute in the 1960s, though there is little information available about those stocks.

Though the Guinea Hog would greatly benefit from additional research and characterization, it is clear that the breed is genetically distinct from improved breeds of pigs and merits conservation. Like other traditional lard-type breeds, however, the Guinea Hog faces great obstacles to its conservation. The pigs do not produce a conventional market carcass, since they are smaller and more fatty than is preferred today. Guinea Hogs are, however, appropriate for use in diversified, sustainable agriculture. They would be an excellent choice where there is need for the services of pigs (such as grazing, rooting, tilling compost and garden soil, and pest control) and also the desire for a small breed. Under such husbandry, Guinea Hogs would thrive, as they always have.

Hereford

Origin: United States
Classification: Standardized breed
Distribution: United States
Status: Critical

The Hereford is a medium size pig breed that is unique to the United States. It was developed in Iowa and Nebraska during the 1920s from Duroc, Chester White, and Poland China bloodlines. Additional breeding and selection led to the identification of 100 animals as foundation stock in 1934, and the National Hereford Hog Record was formed the same year to promote the new breed. Within the first decade of its history, the association attracted 450 members. Most of the interest in the Hereford breed was found in Illinois, Iowa, and Indiana.

The Hereford's name was inspired by its strikingly beautiful color pattern of intense red with white trim, the same as that of Hereford cattle. The breed description calls for hogs to be primarily red, with a white face and two or more white feet. The shade of red can vary, though deep red is preferred. Hereford cattlemen were so keen on the new breed of swine that the Polled Hereford Cattle Registry Association sponsored the formation of the National Hereford Hog Record.

Herefords are adaptable and thrive both in outdoor operations and under confinement systems. They also do well in a wide variety of climates. The pigs are known for their quiet and docile dispositions, making them an excellent choice for young people. The breed is appropriate for 4-H projects because it combines market con-

Hereford Pig

Large Black Pig

formation with a strikingly attractive appearance.

Breeders have emphasized early maturation, and Hereford hogs weigh 200–250 pounds by five to six months of age. Herefords are easy to pasture but also grain-efficient, reaching market weight on less feed than many other breeds. Mature boars weigh about 800 pounds and mature sows about 600 pounds. The sows produce and wean large litters. They make excellent mothers, closely attentive to their bright red and white piglets.

The Hereford began to decline in numbers during the 1960s with the shift away from the commercial use of purebred hogs and toward a three way cross of the Duroc, Hampshire, and Yorkshire breeds. Today, the breed population is estimated at fewer than 2,000 pigs in the United States, most of them found in the upper Midwest and Plains states. The characteristics of the Hereford, however, make it a natural choice for a variety of small scale production systems. If the breed is given opportunity under such systems, it will be able to earn its place in the future.

Large Black

Origin: England
Classification: Standardized breed
Distribution: Britain, with small populations in a few other countries
Status: Critical

The Large Black pig is native to Cornwall, Devon, and Somerset in southwestern England. Originally known as the Lop Eared Black, the breed was selected for large size and efficiency of production on pasture and other forages. The Large Black gained popularity rapidly during the last half of the 1800s, and it was one of the most numerous of the English pig breeds in 1900. A breed association was formed in 1898.

The Large Black was used in small scale production of both pork and bacon. It was also valued for commercial crossing, primarily with the Large White breed (called Yorkshire in the United States). This cross yielded great hybrid vigor, and it was well regarded commercially.

The breed's popularity peaked during the 1920s, and the Large Black was exported to several other countries, including Argentina, Australia,

Brazil, Germany, New Zealand, South Africa, and the United States. After World War II, however, the shift toward intensive husbandry of pigs led to the decline of outdoor breeds which were not competitive indoors. The Large Black nearly became extinct during the 1960s, and it remains one of the rarest British pig breeds. Today, there are about 300 Large Blacks in Britain, with small populations in a few other countries.

The Large Black is best known for its foraging abilities and its maternal qualities. Large Black sows are able to raise and wean large litters of piglets out of doors, and these survival characteristics give it genetic value. As its name implies, the breed is large framed and solid black. Lop ears fall forward over its face, and while they are an impediment to sight, they protect the eyes from damage while the pig is rooting and foraging.

Importations of the Large Black have been made into the United States and Canada at various times, but none has succeeded in making the breed popular. Today, the Large Black is near extinction in North America. It is likely that a reimportation of purebred stock from Britain will be necessary for the breed to have a future here.

Mulefoot

Origin: United States
Classification: Standardized breed
Distribution: United States
Status: Critical

The Mulefoot is an American pig breed that is named for its most distinctive feature, the solid, non-cloven hoof which looks like the hoof of a mule. This characteristic will occasionally occur as a single gene mutation, producing occasional "mulefooted" pigs within a variety of other breeds. In contrast, the Mulefoot breed is consistent in appearance and behavior, with qualities that have made it valuable in American history and a conservation priority today.

The origin of the Mulefoot is unclear, and many theories have arisen about its links with mulefooted stocks in Asia and Europe. The breed is more likely to have descended from the Spanish hogs brought to the Americas beginning in the 1500s. It shares some attributes with the Choctaw hog, and the two breeds likely come from the same ancestral stock, which was loosely selected and managed until the late 1800s.

Mulefoot Pigs

OSSABAW PIGS

By 1900, the Mulefoot had become a standardized breed. It was valued for ease of fattening and production of meat, lard, and especially hams. Mulefoot pigs were distributed throughout the Corn Belt. They were also common along the Mississippi River Valley, where farmers ranged pigs on the islands in the river, putting them out to forage in the spring and collecting them in the fall. Early this century, there were two Mulefoot breed associations and over 200 herds registering purebred pigs.

Mulefoot pigs are compact in appearance and weigh 400–600 pounds. They are solid black with white points occurring rarely. The ears are pricked forward. Some pigs have wattles on either side of the neck, though this is not common. The breed forages well and thrives under extensive husbandry, though its carcass contains too much fat to have much commercial value. This is a major obstacle to its conservation.

The Mulefoot breed is critically rare, with fewer than 100 purebred pigs documented today. Most of these originated in the Holliday herd of Missouri, which is believed to be the last purebred herd in existence. The American Livestock Breeds Conservancy and ALBC member Mark Fields have worked to establish additional breeding herds and to determine the genetic status and purity of the animals which remain.

Ossabaw Island

Origin: United States (Georgia)
Classification: Feral breed
Distribution: United States (Georgia)
Status: Critical

Spanish explorers brought livestock to the Americas beginning in the 1500s, including pigs that escaped or were deliberately set free in the New World. These Spanish pigs were the foundation for the historic populations of pigs, both domestic and feral, in the southern United States. One of these is the Ossabaw, a feral breed which is found on Ossabaw Island, off the coast of Georgia near Savannah.

The Ossabaw pig breed is unusual and important for three reasons. Its history as an isolated island population has meant that the Ossabaw is the closest genetic representative of historic Spanish stocks. Second, the presence of pigs on Ossabaw Island provides scientists with an exceptional opportunity to study a long term feral population which is well documented. Third, the Ossabaw breed is biologically unique, having been shaped by natural selection in a challenging environment known for heat, humidity, and seasonal scarcity of food. Ossabaw pigs may be as small as 100 pounds, but they are able to store astounding amounts of body fat in order to survive during the seasons when there is little to eat. This biochemical adaptation is similar to non-insulin dependent diabetes in humans, making the pigs a natural animal model for this disease. The pigs are also highly tolerant of dietary salt.

Ossabaw pigs are usually black, although some are black with white spots or light with black spots. True Ossabaws do not produce the striped piglets typical of feral hog populations which have had Wild Boar introductions.

Adult pigs are very hairy with heavy bristles on the head neck and topline. The frayed tips of the bristles, a primitive characteristic, is another indication of the distinctiveness of the population. Pigs' snouts are long and slightly dished. Heads and shoulders are heavy, and while they seem out of proportion to the rest of the body, that impression belies the speed and agility of these animals in the dense undergrowth of the island.

Though pigs have lived on Ossabaw for several centuries, they do have an impact on the island's ecology. Environmentalists have become concerned, and some have proposed extermination of the pig population. The island has recently passed from private hands to the state, and the pigs may or may not be allowed to remain, depending upon the state's ability to manage what is considered an ecologically appropriate number of animals. Currently, the size of the population is controlled by hunting, and this practice has its proponents and detractors.

A few populations of Ossabaw hogs have been established on the mainland, most of these at zoos and historic sites. While the Ossabaw makes an interesting exhibit animal, production niches for the breed in agriculture are difficult to imagine. A further complication is the discovery that some pigs on the island test positive for porcine vesicular stomatitis (PSV) and pseudorabies. Though these results only indicate exposure to the disease, pigs may no longer be removed from the island or cross state lines. Thus, the fate of the breed now rests upon the complex questions of how it will be managed in its habitat without having a negative impact on the island's ecology.

Red Wattle

Origin: United States
Classification: Landrace
Distribution: United States
Classification: Critical

Red Wattle pigs are red, wattled pigs that resemble the Duroc in conformation. "Wattles" are fleshy appendages that are attached on each side of the neck. They have no known function, but since they are a single gene characteristic, they will

Red Wattle Pig

usually be seen in crosses with other breeds. The ears are upright with drooping tips.

The origin and history of the Red Wattle breed is obscure, though hogs of this name have long been known throughout the traditional swine production areas of the United States and in Canada. The area of greatest concentration was Texas, and the Red Wattle breed of today probably combines the blood of feral swine from East Texas with several standardized breeds. In the 1980s, three organizations served as registries for Red Wattle pigs, but the breed has never been supported by an active breed association. The current status of the Red Wattle is not known, and more information is needed.

Red Wattle pigs are known for hardiness, foraging activity, and rapid growth rate. They are adapted to a wide range of climates and should be a good choice for consideration in outdoor or pasture-based swine production.

The story of the Red Wattle breed illustrates the problems associated with conservation of regional and local populations. Often poorly documented, even when common, these breeds can rapidly be lost because there is no formal network present to maintain information about them. Collecting information about the history of this breed and documenting its current status are goals for the American Livestock Breeds Conservancy.

Saddleback

Origin: England
Classification: Standardized breed
Distribution: Britain, with small populations in Australia and the United States
Status: Critical

The Saddleback, or British Saddleback, is a hardy and adaptable pig breed known for its ability to thrive out of doors. The name refers to the color pattern of the breed, black with a white band of varying width around the shoulders. The Saddleback's lop ears distinguish it from the similarly patterned Hampshire.

The modern breed originated in 1967 from the combination of two historic breeds of southern England, the Essex and Wessex (sometimes called the Wessex Saddleback). Both of these breeds had been imported to the United States in the early 1800s. The Essex later became extinct

SADDLEBACK PIG

Tamworth Pig

in this country, but the Wessex, a belted breed, became the primary foundation for the American Hampshire breed. Thus, the Wessex, ancestor to both the Hampshire and the Saddleback, was the source of the belted color pattern in both breeds.

Saddleback pigs have traditionally been managed on pasture, and they excel in foraging ability and maternal qualities. They have also been used in commercial crossbreeding systems, especially with Large Whites (called Yorkshires in North America). Today there are about 600 Saddleback pigs in Britain. The breed is slowly increasing in numbers, and it would benefit from a resurgence of the raising of pigs on pasture.

Saddlebacks have been imported into the United States on several occasions, although the stock has generally been absorbed into other breeds and has not been maintained as a pure population. As a result, the number of Saddlebacks in the United States is very small. There is no active breed association nor accurate information about the breed's status. It may be necessary to import breeding stock if the Saddleback is to survive in the United States.

Tamworth

Origin: England
Classification: Standardized breed
Distribution: Australia, Britain, Canada, New Zealand, United States
Status: Rare

The Tamworth originated in central England and takes its name from the village of Tamworth in Staffordshire. Long, lean, and athletic, the Tamworth is probably the most direct descendant of the native pig stock of northern Europe. The breed's genetic distinctiveness and excellent production qualities make it a high conservation priority, both globally and in the United States.

The characteristics of the Tamworth reflect the breed's centuries of selection for an outdoor life. Pigs of this breed were expected to find their own food, especially mast (or acorns) of oak and beech forests. Long, strong legs and sound feet give Tamworth pigs the ability to walk for long distances. The long head and impressive snout make it a good forager. The ginger red coat makes the pigs adaptable to a variety of climates and protect them from sunburn. Sows are prolific, able to produce and care for large litters. Tamworths have an active intelligence, and they are agreeable in disposition. The pigs are medium to large in size, weighing 500–600 pounds.

The Tamworth was traditionally considered a "bacon" breed, meaning that the pigs thrived on low energy foods but grew slowly. They produced meat and bacon that was lean and fine grained. The breed was standardized during the 1800s, becoming uniform in type. A British breed association was established in 1906.

Tamworth pigs were imported to North America beginning in the mid- to late 1800s. The American Tamworth Swine Record Association was founded in 1887. By 1920, the breed was dispersed across most of the United States and Canada, though it never became numerous since it was a bacon breed at a time when lard pigs predominated. More recently, the breed's lack of competitiveness under confinement husbandry caused it to decline. The Tamworth does not readily tolerate the stress of confinement rearing, nor do the pigs put on weight rapidly enough to be competitive under such management.

However, the Tamworth is exquisitely adapted for outdoor production, and the breed merits consideration for any low input systems where thriftiness, hardiness, and reproductive efficiency are desired. The boars are effective breeders, while the sows are successful and attentive mothers. The meat is lean, appropriate for specialty pork niches. The Tamworth is also being used for commercial crossing. It is distantly related to other breeds and will impart hybrid vigor in the offspring.

The Tamworth is rare globally, with a population of 300–400 in Britain and 1,500–2,000 in North America. Though native to Britain, this breed is now most numerous in the United States, and American breeders are pivotal to its global survival. ❖

Chapter 7

Sheep

"Wherever sheep's feet touch the ground, the land turns to gold," says the Spanish proverb describing the value of this versatile species. For thousands of years, sheep have provided people with meat, milk, and wool, and their grazing has been considered an important agricultural service. An appreciation of sheep goes back thousands of years and through many cultures. For example, 5,000-year-old Egyptian carvings show sheep being driven across a newly sown grain field to tread seed into the soil.

The long relationship between sheep and humans has led to the development of a large number of breeds. This diversity has allowed the species to meet a variety of environmental challenges and cultural needs.

Selection for wool quality has been of primary importance in the evolution of sheep breeds. Humans began using wool about 10,000 years ago, first by collecting the fibers shed by wild sheep. After sheep were domesticated, they were selected to produce larger quantities of wool that did not shed out as readily. The lack of shedding meant that all the wool could be harvested at the same time. It also meant that sheep had to be shorn. Today, only the most primitive and unimproved breeds still shed their fleeces.

Wool is a natural, renewable product with a combination of practical and aesthetic characteristics unmatched by synthetic materials. It is strong, durable, elastic, warm, absorbent, and flame resistant. Centuries of selection for wool qualities are reflected in today's extraordinary range of fleece types, textures, and colors.

The characteristics of wool are described in several ways. Fineness of fiber is measured by two methods. The grade, also called the Bradford or wool quality count, estimates the number of skeins of yarn that can be spun from one pound of fleece. The higher the number, the finer the fleece. Fineness is also measured by the micron count, that is the average diameter of the fibers measured in microns. The smaller the number of microns, the finer the fleece. The length of the "staple" (or fiber groupings) in the fleece is also measured, along with its weight. Some fleece fibers are naturally curly and wavy, and this is called "crimp." The

Leicester Longwool Sheep

texture of the wool is referred to as its "handle."

Wool is categorized into four major types: longwool, fine wool, medium wool, and carpet wool. The longwool breeds represent one extreme of fiber selection. Longwool fleeces are loose and open, and they hang in locks. The fibers grow up to twelve inches per year and have a sheen and strength that is imparted to finished products. At the opposite end of the spectrum are the fine wool breeds that produce tightly crimped fibers of very small diameter. These wools are prized for the manufacture of soft, delicate garments. Medium wool falls between longwool and fine wool, and it is useful for many different purposes. In North America, however, most medium wool breeds have been selected for meat production rather than wool quality, so that relatively few of them produce anything other than generic, white wool that is blended in large lots for processing.

Carpet wool breeds are doubled coated, with a coarse, long outer coat for protection from rain, snow, and dust, and a medium to fine inner coat that provides insulation. This type of fleece is unsurpassed for the manufacture of rugs because of its durability. Carpet wool breeds are generally adapted to extreme environments, and the characteristics of the wool are part of that adaptation.

In contrast to the wool sheep just described, "hair sheep" breeds are adapted to hot, humid climates. They lack wool altogether and are covered instead by short hair that is shed annually.

Sheep resources in the United States include all four wool types as well as hair sheep. British breeds have been the primary genetic foundation for the American sheep industry, though breeds from Spain, France, other European countries, and Australia have also been significant. While wool production was historically the primary force in sheep breed development around the world, selection for meat has been of greatest importance in the United States, especially in this century. Sheep dairying has been largely ignored in this country in the past, though interest is increasing.

Spanish explorers and colonists brought the first sheep to the Americas, beginning about 500 years ago. Two remnant populations of these Spanish stocks survive in the United States: the Gulf Coast Native of the Southeast and the Navajo-Churro of the Southwest. English breeds, including the Leicester Longwool and Southdown, were introduced about 1800, along with the Tunis from North Africa. The Spanish Merino, Saxon Merino, and French Rambouillet (a French Merino) were imported beginning in the early 1800s, along with the Cotswold, Cheviot, and several other English breeds. During the middle to late 1800s came the importation of the Dorset, Hampshire, and Suffolk, also from England. Breed imports since 1900 have included the Corriedale from Australia, the Romney from England, the Karakul from the Middle East, and the Barbados and St. Croix hair sheep from the Caribbean. Recent introductions of the Finn sheep from Finland, the Dorper from South Africa, and European dairy breeds have also added to the genetic diversity of the American sheep population.

Despite this genetic breadth, the diversity of sheep breeds in the United States has been neither recognized nor utilized. According to ALBC's *Taking Stock: The North American Livestock Census,* one breed, the Suffolk, accounts for almost 40% of purebred registrations each year. The top four breeds (Suffolk, Dorset, Hampshire, and Rambouillet) account for about 75%. The remaining 25% of sheep registered each year come from 42 other breeds.

Niche production, especially the specialty wool market, has been essential to the survival of rare breeds of sheep. Handspinners and other fiber artisans have a vested interest in diversity and have played a significant role in the conservation of dual-purpose and wool breeds. The use of sheep in sustainable agriculture should increase interest in rare meat breeds, especially those whose parasite and disease resistance, climate adaptation, maternal ability, nonseasonal reproduction, and other characteristics have obvious economic value. It is important that such opportunities be grasped, for without greater use of sheep breed diversity, it will be lost.

Barbados Blackbelly

Origin: Barbados
Classification: Landrace
Distribution: Caribbean Islands, United States
Status: Watch

The Barbados Blackbelly is a hair sheep breed that evolved on the island of Barbados in the Caribbean. The breed descends from hair sheep of West Africa and European wooled breeds that were brought to the Caribbean Islands beginning in the mid-1600s. The Barbados is part of the family of Caribbean Hair sheep breeds attracting attention for their value in a variety of agricultural systems.

Barbados Blackbellies thrive in the hot, humid environments that are challenging for most sheep. Barbados reproduce nonseasonally, meaning that they breed and lamb year-round, and the ewes maintain pregnancies through the heat of the summer. The ewes are prolific, producing twins or triplets. Barbados are excellent foragers and highly resistant to parasites and diseases.

The breed is reddish brown or tan on the body, with black on the belly and legs and black bars on the head. Rams have a mane of thick hair which extends down the neck and brisket and may cover the shoulder. Ears point forward from the side without drooping. Sheep of both sexes are usually polled. Ewes weigh 100–150 pounds, and rams 175–200 pounds. This relatively small body size has been favored by both natural and human selection in the Caribbean. Smaller sheep tolerate heat more easily, and the smaller carcasses are desirable from the farmers' point of view, especially when refrigeration is lacking.

The first documented importation of Barbados sheep into the United States occurred in 1904. The sheep attracted the most interest in Texas, where they have been loosely managed, often crossed with Mouflons (a feral horned sheep) and selected as game animals with large sets of horns. Though this practice has dramatically increased the number of Barbados-type sheep in the United States, it has also made the population genetically variable. For example, most of the Texas sheep are now horned and smaller in size than the original island type.

This lack of genetic consistency has been a major obstacle to the characterization of the Barbados Blackbelly. The breed has also suffered from

Barbados Blackbelly Sheep

the lack of a breed association and registry to maintain pedigree records and promote the sheep for use in agriculture.

In the 1970s, Dr. Lemmuel Goode of North Carolina State University imported a small number of Barbados sheep directly from the island. Sheep from this importation have been the foundation for pure island type Barbados sheep in the United States. Though this type represents a very small portion of the total Barbados sheep population, it is the genetically consistent part of the breed and merits conservation. The number of Barbados sheep in Barbados is not known, but is estimated at a few thousand.

Prospects for the Barbados breed in the United States are improving. The strong market for hair sheep, both here and in other countries, has given encouragement to those who want to maintain the island type of Barbados. Interest in an association and registry for the breed has also increased. The Barbados clearly deserves such attention. Its African origin (unusual in the Americas) and adaptation to tropical climates (unusual for sheep) gives the breed significant genetic value and production importance.

California Variegated Mutant/Romeldale

Origin: United States
Classification: Standardized breed
Distribution: United States
Status: Critical

The Romeldale is an American fine wool breed, and the California Variegated Mutant, or CVM, is its multi-colored derivative. Both the CVM and the Romeldale are unique to the United States and are endangered.

The Romeldale was developed in California by A. T. Spencer early in this century. Spencer purchased the entire contingent of New Zealand Romney rams that were exhibited at the 1915 Pan American Exposition in San Francisco. He bred these rams to his Rambouillet ewes, with the goal of improving both the meat and wool qualities of his stock.

This group of Romney-Rambouillet crosses were bred for several years and selected for both wool and meat quality. They became known as Romeldales. Much of the establishment of the

California Variegated Mutant Sheep

COTSWOLD SHEEP

Romeldale breed was accomplished by the J. K. Sexton family during the 1940s and 1950s. The Sextons selected the sheep for high rates of twinning, maternal ability, and nonseasonal reproduction. Soft-handling wool was also a priority, as was fleece weight (ten to fifteen pounds) and a grade of 60s to 64s. The wool of the Sexton flocks was so highly regarded that for many years the entire clip was sold to Pendleton Mills.

During the 1960s, colored lambs appeared in the Romeldale breed. Glen Eidman, a partner of the Sextons, became interested in these sheep and linebred them for several generations, further selecting for fleece quality. He referred to this group of sheep as California Variegated Mutants, usually shortened to CVM.

Romeldale sheep are white, but the classic color pattern of the CVM is the badgerface, a light body with a dark belly and dark head. This pattern creates a range of shades of color on a single fleece. Selection has increased the range of colors to include gray, black, brown, and moorit. Fleece colors darken during maturation rather than fading as the sheep ages.

CVM and Romeldale sheep may be considered two closely related breeds or two parts of a single breed. With the exception of color, CVMs and Romeldales have similar characteristics. The sheep weigh 150–275 pounds. The rams are active breeders, while the ewes are excellent mothers, prolific and long-lived. Twinning and ease of lambing are considered important breed attributes.

The CVM and Romeldale breeds have never been numerous, and today they are quite rare. The breeds' fleece quality and performance characteristics, however, make them useful for many production systems and valuable to handspinners and other fiber artists.

Cotswold

Origin: England
Classification: Standardized breed
Distribution: Britain, Canada, United States
Status: Rare

The Cotswold is a longwool sheep breed developed on the Cotswold Hills in the west of England. Sheep have been known in this region since the time of the Roman conquest 2,000 years ago, and the Cotswold breed may descend in part from the white sheep brought to England by the Romans.

The Cotswold breed was the foundation for the economic development of the Cotswold region, which is marked by "wool" churches and estates. Wool profits also financed the exploration and trade expansion of the British Empire beginning in the Elizabethan Era to such an extent that even today the British Chancellor of the Exchequer sits on a ceremonial Cotswold wool sack.

The modern form of the Cotswold breed was developed between 1780 and 1820, and importations to the United States began about that time. The American Cotswold Record Association was established in 1878. The breed was particularly valued for crossing onto fine wool ewes to produce large lambs that could be used for meat or kept for their heavy fleeces. This success in crossbreeding, coupled with the lack of market demand for pure stock (thought to be too large and slow growing) led to the near extinction of the Cotswold by the middle of this century. In the last twenty years, however, the breed has experienced a revival of interest, due primarily to the rediscovery of its wool by fiber crafters.

Cotswold fleece hangs in locks, and the fibers have a lustrous, silky sheen. A heavy forelock falling over the face is a characteristic of the breed. The fleece has a ten- to thirteen-inch staple length and weighs thirteen to fifteen pounds. Cotswold wool encompasses the greatest range of grades (38 to 48) found in the longwool breeds.

Cotswolds are large-framed sheep, and mature animals weigh 200–300 pounds. They are efficient grazers, but require good forage for best performance. Cotswold sheep are known for their docile dispositions, and the ewes are excellent mothers. The breed's name reflects the traditional habitat of the sheep: "cots" for the cotes or small sheep folds where they were sheltered at night on the "wold," treeless hilly land.

Until recently, only white sheep were considered part of the Cotswold breed, though registration is now available for black sheep. These sheep are called black but their colors vary across a beautiful range of silver, bluish gray, and charcoal hues.

Dorset Horn

Origin: England
Classification: Standardized breed
Distribution: International
Status: Watch

The Dorset is an ancient breed likely developed

Dorset Horn Sheep

from white faced, horned, short-wooled sheep that thrived in the sheltered valleys and lush hill pastures of southwestern England. Today, the Dorset is globally distributed and found in two varieties, Horned and Polled. The Dorset Horn (or Horned Dorset) is of conservation interest.

Dorsets are white sheep with open faces and long bodies. Ewes weigh 150–200 pounds, and rams weigh 225–275 pounds. The sheep are avid grazers and use feed well. Ewes are heavy milkers, easily able to raise their own lambs. They have also been used in dairying. Dorset sheep produce between five and nine pounds of medium grade wool per year. In the Horn variety, both ewes and rams carry horns. Ewes' horns are light, curving forward neatly; rams' horns are heavy and spiral out as well as curving forward.

The single most important attribute of the Dorset is its extended breeding season, also called nonseasonal or aseasonal breeding. This characteristic, not found in any other British breed, is an important performance trait. Sheep of most breeds are seasonal breeders, mating in the fall and producing lambs in the spring. In contrast, Dorsets can be bred in the spring for production of lambs in the fall. Some ewes will raise two sets of lambs a year, with multiple births not uncommon. Dorsets tolerate heat well, and heat tolerance contributes to Dorset rams' ability to breed earlier in the season than rams of other breeds.

Dorset Horn sheep were imported into the United States about 1860, though the breed did not appear in large numbers until the 1880s. The Continental Dorset Club was organized in 1890. In the early 1950s, a genetic mutation causing polledness occurred in the flock at North Carolina State University. The Polled Dorset variety was developed there. (Polled Dorsets were also developed in Australia earlier in this century, though they originated from the introduction of Corriedale and Ryeland blood into the Dorset Horn.)

The first Polled Dorsets were registered by the Continental Dorset Club in 1956, and they have been included in the same flock book with Dorset Horns. The success of the polled variety has caused the Dorset to become the second most popular sheep breed in the United States. Polled Dorset sheep have almost completely eclipsed Dorset Horns, and the latter is experiencing a precipitous drop in registrations.

The Polled Dorset and Dorset Horn have diverged somewhat over the past 40 years. Though differences between the populations have not been documented, anecdotal evidence suggests that the intensive selection of Polled Dorsets for early maturation and rapid growth rate may have been made at the expense of selection for nonseasonal breeding, which has declined as a result. Given the possible impact of these changes and the long divergence of American Dorset Horn sheep from other Dorset populations, the Dorset Horn has been included as a conservation priority.

Gulf Coast Native

Origin: United States
Classification: Landrace
Distribution: United States (Southeast)
Status: Critical

Gulf Coast Native sheep descend from the Spanish flocks brought to the New World by explorers and settlers beginning in the 1500s. Spanish sheep in the Southeast were shaped primarily by natural selection, becoming well adapted to the heat and humidity of the environment. Though the word "Native" in the breed's name is not precisely correct, the sheep were so thoroughly transformed in the New World that a new breed was created.

The genetic origins of the Gulf Coast Native breed are not known, since a wide variety of types and breeds of sheep existed in Spain at the time. Churro sheep, multi-purpose animals used for meat, milk, and coarse wool, were commonly brought to the Americas by the Spaniards, however, and may well have contributed to the breed's foundation. At the same time, the Gulf Coast Native's fine wool suggests a contribution from pre-Merino types as well.

Gulf Coast Native sheep were used across the Southeast by Spanish missionaries, Native

Gulf Coast Native Sheep

Americans, and European settlers as far north as the Carolinas. These sheep fit their challenging environment so well that for centuries they were the only sheep to be found in the deep South, providing wool and meat for home production. The development of anti-parasite medications in this century allowed the introduction of other, more productive sheep breeds in the Southeast, and Native sheep were discarded by most farmers. The breed was saved only through the action of a few Southern families.

Gulf Coast Native sheep lack wool on their faces, necks, legs, and bellies, an adaptation to the heat and humidity of the South. Otherwise they tend to vary somewhat in aspects of physical appearance. Variability has also resulted from the isolation of different strains of the breed. While most sheep are white, blacks and browns also occur, and many individuals have spotted faces and legs. Most rams and some ewes are horned, although both sexes may also be polled. Native sheep vary in size, with rams weighing 125–200 pounds and ewes 90–160 pounds.

What is consistent among Gulf Coast Native sheep is their exquisite adaptation to an environment that is generally difficult for sheep. Gulf Coast Natives breed and lamb year-round. The ewes make excellent mothers, pasture lambing without assistance. Ewes usually produce one lamb at a time, though the lambs are vigorous and grow rapidly. Gulf Coast Native sheep have well-documented resistance to gut parasites and to footrot and other diseases. These valuable performance characteristics recommend the breed for low-input sheep production.

Like other landrace breeds, the Gulf Coast Native has suffered from a lack of documentation and description. Researchers at the University of Florida and at Louisiana State University have addressed this problem by working to locate breed populations and describe the breed's characteristics. Sheep in other states have received less attention, but efforts are now underway to inventory and register Gulf Coast Native sheep across the Southeast. A breeders' association has been formed to promote the breed, and a breed registry is held by the American Livestock Breeds Conservancy.

Hog Island

Origin: United States
Classification: Feral
Distribution: United States (Virginia)
Status: Critical

Hog Island is a barrier island near the mouth of the James River in Virginia. About 200 years ago, a flock of sheep was established on the island using locally available sheep of British origin. The sheep of Hog Island evolved in response to the island's natural selection for hardiness, foraging ability, and reproductive efficiency. Some introduction of outside blood occurred, most recently a Hampshire ram in 1953, but the isolation and selection of the habitat has shaped the population into a distinct breed.

Hog Island sheep are one of the few populations of feral sheep in the United States. Feral sheep are rare worldwide, because sheep do not easily adapt to unmanaged habitats. They occur most often on islands which lack predators.

Hog Island sheep vary in physical appearance. Most of the sheep are white with spotted faces and legs, though about ten percent are black. Newborn lambs are frequently spotted over the body, but the spots usually disappear as the lambs mature. Wool is of medium weight and variable in type and amount. Both ewes and rams may be horned or polled. Mature animals weigh 125–200 pounds.

The entire sheep population was removed from Hog Island during the 1970s when the island was purchased by The Nature Conservancy. A remnant flock of the breed is being kept by Gunston Hall, Mt. Vernon, and a few other historic sites in Virginia to portray eighteenth century sheep raising. Domestic husbandry of Hog Island sheep will change some of the breed's characteristics, but it can save what remains of this critically rare breed.

Hog Island Sheep

Jacob (American)

Origin: England
Classification: Standardized breed
Distribution: Canada and United States, with related strains in Britain and other countries
Status: Rare

Sheep with spots have been described in many cultures throughout history, appearing in works of art from the Far East, Middle East, and Mediterranean regions. Among these accounts is the Biblical story of Jacob, who bred spotted sheep and for whom this breed is named.

Spotted sheep were documented in England by the 1600s and were widespread by the mid-1700s. They became popular in England as ornamental, or "park" sheep. Jacobs were ideal for this role, as they were picturesque but required minimal care. Scant selection occurred for anything but hardiness, spots, and four horns. The result was a primitive breed that looked after itself well.

Jacobs are small, horned, black and white sheep. Ewes weigh 80–120 pounds, and rams 120–180 pounds. The sheep are white with colored spots or patches. The colored portions of the fleece are usually black, but they can also be brownish or a lighter color called lilac. The Jacob is a multi-horned or "polycerate" breed. Most animals have two or four horns, though six horns also occur. Both sexes are horned, and the rams can have horns of impressive size and shape.

The breed produces a medium fleece that is light and open, with a staple length of four to six inches and a weight of three to six pounds. Unlike most other medium wool breeds, quality of the fleece has been a major selection factor in the recent history of the Jacob breed. As a result, it is much sought after by fiber artisans, who enjoy its characteristics and color combinations – black, white, or a blend of the two.

Jacob sheep were first imported into North America beginning in the mid-1900s, and most of today's population descends from imports of the past 30 years. The breed has enjoyed widespread popularity among small flock holders as well as handpsinners and weavers. North American breeders have selected primarily for fleece characteristics, and the conformation of the sheep has remained very much like its historical description. Variability is present, but this is characteristic of an unimproved, primitive breed. During the same period of time, the British Jacob has been selected for greater commercial productivity, including larger size and more uniform appearance. In this way, the populations in Britain and North America have

Jacob Sheep

Karakul Sheep

diverged. The American Livestock Breeds Conservancy has listed the North American population of Jacobs as a conservation priority.

The distinctive appearance of the Jacob has sometimes worked in its favor but at other times has proven an obstacle to its conservation as a pure breed. Spotted sheep of all shapes and sizes, including spotted Jacob-Dorset and other crosses, have been sold as Jacobs to unsuspecting buyers. Identifying and recording the purebred Jacob population has been a continuing challenge for American breeders.

Karakul (American)

Origin: United States
Classification: Standardized breed
Distribution: Canada and United States, with related strains in Central Asia
Status: Rare

The Karakul breed originated in Central Asia, and its name comes from the village of Kara Kul (or Black Lake), which lies in the Bokhara region between Turkmenistan and Afghanistan near the Caspian and Black Seas. Karakuls were historically used for the production of meat, fat, wool, and pelts. Of most value were the pelts of lambs, called Persian lambskins or *astrakhan*, which were used to make a variety of warm, soft, and elegant garments.

Karakul sheep were introduced into the United States between 1909 and 1936. The population in North America has been genetically divergent from Central Asian stocks since the time of importation, and this warrants classification of the American Karakul as a separate breed. Karakuls were imported to establish a pelt industry in the United States, but this effort was not successful, and the breed has always been relatively rare.

The Karakul is a member of the family of fat-rumped and broad-tailed sheep. This breed family originated in Central Asia and Northern Africa, and it is characterized by the deposit of fat at the base of the tail instead of throughout the body. This fat is distinctive in texture and flavor

KATAHDIN SHEEP

from other body fat and is highly valued in the cuisine of Central Asia. The rest of the carcass is very lean.

The soft birthcoat of the Karakul matures into a fleece of carpet wool, which consists of a long, lustrous outer coat and a fine, soft inner coat. Adding interest to this unusual fleece is the breed's array of colors. Most sheep are born black and fade to gray with age, but red, tan, gold, and brown also occur. Handpsinners and weavers provide the market for most of the Karakul wool produced in this country. Fleeces weigh five to ten pounds and have a staple length of six to twelve inches.

Though used primarily for the production of wool, Karakuls are multi-purpose sheep with many valuable characteristics. They are hardy and adaptable, thriving under rugged conditions in a variety of climates. The sheep are aggressive grazers and browsers, useful wherever pastures need improvement. Karakuls are generally seasonal breeders, producing a single lamb. Ewes are attentive mothers that milk very well. Karakuls are medium in size, weighing 125–175 pounds. Rams are horned and ewes are polled.

Katahdin

Origin: United States
Classification: Standardized breed
Distribution: Canada, United States
Status: Watch

The Katahdin is a hair sheep breed which was developed at Piel Farm in Maine beginning in the 1950s. The goal was to produce a hardy meat sheep that did not require shearing. Virgin Islands hair sheep contributed the hair coat, as well as hardiness and prolificacy. The Suffolk and other wooled breeds were used to improve size and carcass quality. During the 1970s, the Wiltshire Horn, a hair sheep from England, was also incorporated to a limited degree in order to increase size and improve carcass quality. The resulting breed was named for Maine's Mt. Katahdin.

Katahdin sheep are medium in size, with ewes weighing 120–160 pounds and rams weighing 180–250 pounds. As hair sheep, Katahdins shed without being sheared. The sheep vary somewhat in the type and amount of fiber in their coats, though it is preferred that sheep be completely free of woolly fibers after shedding. The coat can be any color or pattern. Most of the sheep are polled, although horned animals sometimes occur in the breed.

The Katahdin is selected to be an efficient meat producing sheep adapted to a wide variety of environments. While the winter coat provides enough protection for the sheep to thrive in cold climates, their short, hair coat allows them to tolerate the heat and humidity of warmer regions. Katahdins demonstrate greater parasite resistance than commercial wooled breeds, and research is underway to document other performance characteristics.

In 1986, Katahdin Hair Sheep International (KHSI) was formed to register sheep, record performance information, and promote the breed. The Katahdin is increasing in numbers in North America and has also been exported to Central and South America. The breed is taking advantage of the fundamental shift in the American sheep industry caused by the decline of the wool market. This success story is an excellent illustration of the principle that breed resources are conserved so that they can be available when production goals and management systems change.

Leicester Longwool

Origin: England
Classification: Standardized breed
Distribution: Australia, Britain, New Zealand, United States
Status: Rare

The Leicester Longwool is one of the "luster longwool" breeds, so designated for the sheen and brilliance of their wool. The sheep appear to shine just after shearing, when the clean wool next to their skin catches the sunlight and makes them glisten for a few days before the dust and dirt of their environment catches up to them and the glow is hidden for another year.

The Leicester Longwool breed is also known as the English Leicester (pronounced *lester*). The breed was developed in England in the mid-1700s by innovative breeder Robert Bakewell, the first to use modern selection techniques to improve livestock breeds. Bakewell transformed a coarse, large boned, slow-growing animal into one that grew rapidly for market and produced a higher quality fleece.

LEICESTER LONGWOOL SHEEP

News of Bakewell's ideas reached the colonies before the American Revolution and so intrigued George Washington that he made reference to them in several letters. Washington was particularly interested in Bakewell's sheep, writing that he made the "choice of good rams from the English Leicester breed" for his own flock. In 1837 the agriculturist Youatt wrote that "within little more than half a century the New Leicester had spread themselves to every part of the United Kingdom and to Europe and America."

The Leicester Longwool was highly prized in America, especially for its use in crossbreeding to improve "native" stock. During the 1800s, however, the breed lost favor to the Merino and other fine wool breeds. After 1900, the Leicester Longwool fell into decline and was likely extinct in the United States during the 1930s or 1940s. A very small population remained in Canada. In 1990, Colonial Williamsburg Foundation, a historic site in Virginia, reestablished the breed in North America by importing sheep from Australia. Several conservation flocks have now been established, and the population of Leicester Longwool sheep in North America is increasing. This is important, given that the breed remains rare globally.

Leicester Longwools are medium to large sheep, weighing 180–250 pounds. The fleece is heavy, curly, soft-handling, and lustrous with a spiral-tipped staple up to eight inches. Fleeces weigh from eleven to fifteen pounds, occasionally up to twenty pounds. Leicesters are eager grazers, making good use of abundant pasture. When mixed flocks of Merinos and Leicesters are driven along road sides in Australia, all of the Merinos have their heads up, watching what is going on, while the Leicesters are busy with their heads down, chomping down the succulent roadside grasses. Leicesters are docile and easy to handle, but they do not care for herding dogs. Herding with dogs is likely to result in the whole flock proceeding to the barn backwards – facing down the dog!

The Leicester Longwool has been of great historic and genetic value, having a part in the founding or improving of many other breeds, including the Border Leicester and the Corriedale. While distinguished by its past, this breed's future is far from secure, and it is a conservation priority.

Navajo-Churro

Origin: United States (Southwest)
Classification: Landrace
Distribution: United States
Status: Rare

America's first domestic sheep were small, rugged Churro sheep from Spain brought by Francisco Coronado in 1540 and Don Juan de Onate in 1598. Sheep were used as a source of meat for the explorers and for the missionaries who followed and established a chain of missions throughout the region that is now Mexico and the southwestern U.S.

During the "Golden Age" of New Mexico's sheep industry (1788–1846), master weavers from Mexico promoted their skills, and trade in textiles and livestock was significant. In 1807, Zebulon Pike reported that a single flock of Spanish sheep could number 20,000. Southwestern churros went down the Chihuahua and the Santa Fe trails and eventually all the way to California.

Churro sheep also found their way into the hands of Native Americans. The people used sheep for meat but especially came to value the wool, a wonderful new material that was quickly adopted for production of textiles. Sheep quickly became the basis for subsistence and trade in Hispanic and Native American economies. Selection for fleece character, coupled with the natural selection of the challenging environment of the arid Southwest, forged the Navajo-Churro breed.

During the 1860s, the Navajo-Churro sheep population was nearly destroyed as part of the United States government's efforts to subjugate the Navajo people. Churro stock was further diluted by continuing efforts to "improve" Native

Navajo-Churro Sheep

American flocks through introduction of other breeds. The government's attempts to control rangeland erosion led to further slaughter of Native American flocks. By the mid-1930s, very few "purebred" Navajo-Churros remained, and they were scattered across the Southwest.

In 1977, Dr. Lyle McNeal and other conservationists began an effort to protect the breed from extinction. McNeal, an animal scientist at Utah State University, founded the Navajo Sheep Project to conserve the Navajo-Churro, especially among the traditional herders and weavers of the Southwest. The Navajo-Churro Sheep Association and registry were formed in 1986 and include the many groups of people who have historically been involved with the breed. The Navajo-Churro has benefited from this broadly based conservation effort, and registrations have risen steadily in the last decade. The Navajo-Churro is still rare, but its survival now seems assured.

The Navajo-Churro sheep is a hardy breed, adapted to the adverse conditions found in hot, dry deserts and sub-zero climates. It has long legs, a narrow body, and light bones. Both ewes and rams may be polled or have two or four horns. Mature rams average 160–200 pounds and mature ewes about 100–120 pounds. The breed is noted for prolificacy. Lambs are easily born and vigorous, and the ewes are excellent mothers. The weaning rate is high, especially when the environment is considered.

The wool of the Navajo-Churro is considered a carpet wool, and it is used primarily for rug weaving. Sheep have a double coated fleece that weighs four to six pounds. The fine, soft inner coat provides insulation, and the long, coarse outer coat protects the inner coat from dust and dirt while repelling rain and snow. The fleece is low in grease, which makes it easy to process. Navajo-Churro sheep are found in several colors, including white, silver, blue, brown, red, black, and spotted. Patterns such as badger face are common. This range of hues is valued by wool crafters and sheep breeders alike.

Oxford

Origin: England
Classification: Standardized breed
Distribution: International
Status: Watch

The Oxford breed is named for its home region of Oxfordshire, England. It was developed in the 1830s through the crossing of Hampshires from the county to the east with Cotswolds from the west. The resulting sheep combined the hardiness, muscling and wool quality of the Cotswold with the carcass quality and grazing ability of the Hampshire. The breed was standardized during the 1850s, and by the 1860s was known by the name Oxford.

The Oxford is the largest of the English "down" breeds, developed for the lush pastures on the hills, or downs, of southern England. Other down breeds include the Hampshire, Shropshire, and Southdown. Oxfords were selected to produce meat and wool from abundant feed, and as a result they do not prosper on regimens of sparse feed and marginal management. They are nonetheless rugged and tolerate harsh climates.

Oxford sheep were imported into North America in the late 1840s, with importations continuing into the 1900s. The Oxford Down Record Association was founded in 1882 at Xenia, Ohio. The breed enjoyed its greatest success in eastern Canada, the Great Lakes states, and the Cornbelt, where Oxford rams were widely used on commercial ewes to improve the growth rate and carcass quality of lambs.

Oxfords are large, with rams weighing about 300 pounds and ewes 200 pounds. The sheep are quiet, docile, and easy to manage. Ewes produce single or twin lambs and show good maternal ability. While the "down" breeds are typically blackfaced, the Oxford is more variable in color on its face and legs, which may be one of several shades of brown or gray. Oxfords produce eight to twelve pounds of medium white wool that grades 50–54 and has a staple length of three to five inches.

The Oxford has historically been one of the most important of the terminal sire breeds, used as the final cross onto commercial ewes for production of fast-growing market lambs. The breed enjoys modest popularity in other parts of the world, especially in Britain, where, after being rare for some time, it is once again increasing in numbers. The Oxford remains rare in the United States, having been supplanted by the Suffolk. This situation is difficult to explain, as research has shown time and again that the Oxford excels as the sire of market lambs and the breed's overall profitability may be second to none.

Oxford Sheep

Santa Cruz Sheep

Santa Cruz

Origin: United States (California)
Classification: Feral
Distribution: United States (California)
Status: Critical

Santa Cruz sheep are named for Santa Cruz Island, one of the Channel Islands off the coast of California. Sheep have been on the island for at least 70 years and perhaps as long as 200 years. Local lore suggests that sheep were moved onto the island in the early 1800s. Ranching operations on the island included sheep, possibly Merino and Rambouillet types. Some of these animals escaped and a feral population was established. Over time, the Santa Cruz breed developed, through the geographic isolation of the population and the pressures of natural selection.

In 1978 The Nature Conservancy (TNC) acquired ownership of nearly all of Santa Cruz Island. The feral sheep population had overgrazed the island and threatened the survival of indigenous vegetation. TNC began a sheep eradication program in 1980 in an effort to save the island ecosystem from further destruction.

The American Livestock Breeds Conservancy convinced TNC of the importance of the feral sheep, and the two organizations agreed to cooperate in a rescue effort to conserve a portion of the population. In 1988, twelve lambs were brought off the island by a team of Nature Conservancy and ALBC volunteers. The sheep were placed with five California breeders and a conservation effort was begun. ALBC member Earl Crews brought additional animals off of the island in 1991. The mainland population has now increased to about 125.

Like many other island-adapted animals, the Santa Cruz sheep are small in size. Most of the sheep are white, but black, brown, and spotted sheep are also found in the population. The fleece is fine to medium and very soft. Sheep have little or no wool on their bellies, faces, and legs, and many have short, woolless "rat" tails. Santa Cruz sheep are exceptionally hardy, with virtually no birthing problems, a high survival rate, and an ability to thrive on marginal forage.

The Santa Cruz Island sheep breed is an important genetic resource. Its historic background, long period of isolation, and adaptation to a challenging environment have given the breed an array of characteristics not found among commercial breeds. Conservation outside of its native environment is not ideal, because changes in

SHETLAND SHEEP

selection pressure will gradually change characteristics, but it provides the only opportunity for the breed to survive. The American Livestock Breeds Conservancy has begun a conservation program for the sheep, including the collection of semen from the original island-caught rams. Semen can be used in the future if needed to re-establish original traits.

Conservation of the Santa Cruz sheep breed will require the participation of several dedicated breeders who are able to manage flock-based breeding programs and carefully observe the sheep's characteristics. Though this type of conservation effort is not for everyone, it does have the satisfaction of protecting a truly unique genetic resource from disappearance.

Shetland

Origin: Scotland (Shetland Islands)
Classification: Landrace on Shetland Islands, Standardized breed elsewhere
Distribution: International
Status: Watch

Sheep have lived on the Shetland Islands for well over 1,000 years, adapting to the harsh environment and thriving in the cold, wet climate. The sheep of Shetland were an important part of subsistence agriculture of the islands, and the rugged habitat and geographical isolation produced a breed that is distinct and significant.

The Shetland breed likely descends from ancient Scandinavian sheep, and it is a member of the northern short-tailed sheep breed family. Historically, only a few Shetland sheep were exported, and it was not until recently that large populations were established on the British mainland and in other countries. Though fleece continues to be the breed's primary product today, Shetlands in Britain are also finding a commercial niche for crossing with Cheviots and other breeds to produce market lambs.

Shetland sheep are fine boned and small in size. Rams weigh 90–125 pounds, and ewes weigh 75–100 pounds. Most rams have spiraled horns, while most ewes are polled. Shetland sheep are calm and charming in disposition, docile, and intelligent.

The Shetland breed is especially prized for its wool, which is fine, soft, and strong. Fleeces average two to four pounds and vary in crimp from wavy to straight. Other characteristics of the fleece vary according to recent selection history. Populations of Shetlands in Britain, for example, have been selected for more standardized characteristics. These sheep tend to be single coated with fiber diameter averages of 23 microns and staple lengths of two to five inches. Landrace populations, such as those on the island of Foula, include a greater range of fleece types. These sheep may be double coated, with coarser outer wool of 30-40 microns and finer inner coat wool of 12-20 microns. Eleven colors and thirty color patterns are recognized in the Shetland breed. This diversity is a great asset both to the breed and to the fiber artisans who enjoy using its fleeces.

A few importations of Shetland sheep are documented in North America during the past two centuries. For example, Thomas Jefferson, owned a small flock of Shetland sheep at Monticello. None of the historic flocks, however, survived as purebred populations. Most Shetland sheep in North America descend from a 1980 importation of 32 sheep by the late G.D. Dailley of Ontario, Canada. The North American Shetland Sheep Registry began keeping a North American flockbook in 1991.

The North American Shetland Sheep-breeders Association was formed in 1994 to promote the breed and to support the growing number of Shetland breeders in North America. The Shetland breed has prospered in recent years to the extent that it is no longer considered endangered by the Rare Breeds Survival Trust in Britain. Despite this success, there are concerns about the loss of genetic diversity within the breed. For example, white sheep now predominate on the British mainland and several of the color varieties have become rare. Though the Shetland sheep seems to have a secure future, North American breeders have an important role to play, not only through breed promotion on this continent, but also in the conservation of its entire range of colors, markings, fleece types, and other characteristics.

St. Croix

Origin: United States
Classification: Standardized breed
Distribution: United States, with a small population in Canada
Status: Rare

The St. Croix is an American sheep breed that is part of the Caribbean Hair sheep family of breeds. Caribbean Hair sheep were developed from the hair sheep of West Africa and a few European wooled sheep that were brought to the Caribbean beginning in the 1600s. The sheep proliferated as subsistence livestock, and they were also valued for the manure critical to sugar cane production.

Over time, Caribbean Hair sheep became well adapted to the heat and humidity of their environment. The hair coat, which eliminates the need for shearing, is part of this adaptation. Today, there are several landrace populations within this

St. Croix Sheep

breed family in the Caribbean. Two breeds, the Barbados and the St. Croix, are also found in North America.

In 1975, Dr. Warren Foote of Utah State University imported 22 ewes and three rams from St. Croix in the Virgin Islands. Foote chose a relatively consistent group of polled, white sheep from a landrace called Virgin Islands White. Offspring of this imported group were further selected for consistency of conformation, and this process has resulted in the development of the St. Croix, a standardized breed in the United States.

The St. Croix is adapted to the heat and humidity of a tropical climate, and this adaptation has several manifestations. The breed has well-documented parasite resistance, far superior to that found in most other sheep breeds. It is small, with ewes averaging 120 pounds and rams 165 pounds. St. Croix sheep are known for high fertility, and ewe lambs become fertile at about six months of age. Ewes often produce twins and have plenty of milk to raise them. Two lambings a year are not uncommon.

St. Croix sheep are excellent foragers and very easy keepers. Their browsing ability makes them useful for land management, including mowing grass in orchards and the control of invasive pest plants. Though heat-tolerant, the sheep can be raised in many parts of North America. In colder areas, they grow a heavy winter coat of wool and hair that is shed in the spring. This combination of characteristics makes the breed an excellent choice for low-input meat production. The St. Croix is increasing in numbers, and though it is still rare, the breed's future seems promising.

Tunis

Origin: United States
Classification: Standardized breed
Distribution: Canada, United States
Status: Rare

North African sheep, variously described as "fat-tailed," "broad-tailed," and "Barbary" sheep, were imported to the United States during the late 1700s. References to these sheep appear in letters, journals, and farm records of some of the leading agriculturists and citizens of the day, including John Adams, Thomas Jefferson, Richard Peters, and George Washington. Gradually, a uniquely American breed was created. The breed name Tunis described its connections to foundation stock from Tunisia in North Africa.

Tunis sheep spread throughout the mid-Atlantic and southeastern United States and were

Tunis Sheep

well adapted to the heat and humidity of these regions. The Tunis was the mainstay of sheep production in the upper South and mid-Atlantic regions until the Civil War, when nearly all Southern stock was destroyed. After the war the Great Lakes region and New England became strongholds for the breed. It is only recently that the Tunis has again been seen in the Southeast.

Tunis are striking in appearance, with red faces and legs and ivory colored fleeces. Their clean heads and lop ears are distinctive, and the breed gives the impression of activity and intelligence. The sheep are medium in size, weighing 150–200 pounds. Both sexes are polled. Lambs are born red, gaining a lighter fleece with maturity, but retaining red legs, heads, and ears.

The Tunis is a dual-purpose breed which has been selected primarily for meat production. Market lambs are economical to raise and produce carcasses of excellent quality. The fleece is a medium grade wool which turns white during processing. Fleeces weigh seven to twelve pounds with a staple length of three to five inches. Ewes are prolific and fertile, consistently producing twins. They are heavy milkers and make attentive mothers. Tunis sheep are good grazers and easy keepers, appropriate for forage-based production.

The recent history of the Tunis is less well described than its early years. The breed's lack of documentation in performance traits has limited its commercial appeal. Fortunately, the number of Tunis sheep is now increasing, and the breed has the potential to become more widely used for low-input production of meat, wool, and perhaps milk. Such promotion would aid in the conservation of the Tunis, one of the few sheep breeds unique to North America.

Wiltshire Horn

Origin: England
Classification: Standardized breed
Distribution: Australia, Britain, New Zealand, United States
Status: Rare

The Wiltshire Horn is native to southwestern England, descending from the large, white sheep that grazed on Salisbury Plain for centuries. It gets its name from the county of Wiltshire. The breed is distinguished by its self-shedding hair coat,

Wiltshire Horn Sheep

making it the only British hair sheep breed.

Wiltshire Horn sheep were traditionally used for the production of mutton and also selected for use as terminal sires for commercial lamb production. When bred to commercial wooled ewes, Wiltshire rams impart both size and hybrid vigor to offspring.

The Wiltshire was the dominant type of sheep across southern England during the 1700s. It became rare after 1800 when the change to higher stocking rates and more intensive husbandry favored the newly developed Southdown breed. The Wiltshire Horn found favor elsewhere in England, at first in the eastern counties and the Midlands, and then in Northamptonshire. During this century, it has become established in northern Wales, especially Anglesey.

The sheep are large in size, with mature rams reaching weights of 300 pounds or more. The distinctive coat consists of a mixture of wool and fine hair which is shed in the spring. Both males and females have impressive spiral horns. Lamb crops average about 160 percent. The breed is hardy with excellent maternal ability, but depends upon high quality nutrition and good management to meet its potential.

Wiltshire Horn sheep were exported from Britain to several other countries during the 1950s and 1960s, including Australia and New Zealand, Kenya, Venezuela, Argentina, and France. Only in Australia and New Zealand, however, were large purebred populations maintained. This is not surprising given that the breed's purpose was to be used in crossbreeding. Today the Wiltshire Horn is globally rare, though it is slowly increasing in numbers.

Wiltshire Horn sheep were first documented in North America in 1624 at Plimoth Plantation. The breed persisted for some time, though it was not numerous and was never well documented in the historical record. The breed has never had an association formed on its behalf.

During the 1970s, Wiltshire Horns were imported to the United States from Australia and used by Piel Farm in Maine in the development of a modern hair sheep breed called the Katahdin. In the 1990s, Wiltshire Horn sheep were again imported to the United States, and these imports have greatly increased the genetic breadth of the American population. Semen collection and the use of several ram lines may revive the American population so that the breed can take advantage of the dramatic growth of interest in hair sheep.❖

Chapter 8

Conservation Breeding

"What we can hand over after our time of work is not just what we have managed to add to the heritage; it is the whole heritage with the little we have managed to add."

–Dag Hammarskjold

DOMESTIC ANIMALS closely share a habitat with people. Unlike wild species, these are working animals and they must continue to be useful in order to survive. Cary Fowler and Pat Mooney found this to be true of all agriculture, writing that diversity in animals and plants has to be used in order to be valued enough to be conserved.

Conservation of rare breeds has always been done by individual farmers and animal breeders. In fact, almost all the breeds in this book have at some point been close to extinction only to be saved by a few stubborn people who simply refused to give them up, regardless of the fashions or opinions of the times. In few other areas of work can an indvidual's accomplishments be so significant.

Conservation of rare breeds also depends upon the existence of several herds within each breed. Decentralization protects the breed against the sudden, devastating loss to disease or predation that can occur to a single group of any size. Multiple herds also insure against the insidious losses that can occur gradually if animals are held in a single herd and subjected to a single selection program. A network of several herds – each independently managed and selected – is the foundation for breed conservation.

You can become a part of the effort to save rare breeds from extinction. Conservation is challenging, but it does not have to be expensive nor take a lot of space. The first step is to learn everything you can about the characteristics and management of the species and breeds of interest. The second step is to become acquainted with other breeders who have experience with the breed. The third step is to develop a philosophy and goals that fit the breed and can guide the selection of foundation stock and the direction of a breeding plan.

Conservation breeding is the subject of *A Conservation Breeding Handbook,* written by D. P.

MAN HOLDING TAMWORTH PIGLETS

Sponenberg and C. J. Christman, published by ALBC in 1995. The *Handbook* describes the entire process, from selection of a species and breed and purchase of foundation animals to the design of a breeding protocol for a critically endangered population. This book is highly recommended for people who want to become rare breeds stewards, and it is available from the American Livestock Breeds Conservancy.

The American Livestock Breeds Conservancy

The American Livestock Breeds Conservancy (ALBC) is a nonprofit organization protecting genetic diversity in American livestock and poultry through the conservation of nearly 100 endangered breeds of cattle, donkeys, goats, horses, sheep, swine, chickens, ducks, geese, and turkeys.

The American Livestock Breeds Conservancy was founded in New England in 1977. Agricultural historians, seeking historically authentic livestock for interpretive programs at Old Sturbridge Village and other historic sites, discovered that many of these breeds were nearly extinct. The historians joined with farmers and scientists to form the American Minor Breeds Conservancy (AMBC) to conserve endangered livestock breeds. The organization moved to Pittsboro, North Carolina, in 1985. Its name was changed to the American Livestock Breeds Conservancy in 1993 in order to reflect the growing threat to all purebred livestock and poultry breeds and a widening of conservation activities.

The American Livestock Breeds Conservancy does research on breed status and characteristics, and this research also determines conservation priorities. Programs include the operation of a gene bank; bloodtyping and DNA analysis for breed characterization; rescues of threatened populations; and the development of genetic recovery breeding protocols. ALBC educates the public and policy makers about the importance of conserving genetic resources and provides technical support on conservation breeding, registry operation, and livestock use for individuals, breed associations, and agricultural organizations.

The American Livestock Breeds Conservancy was chartered in Vermont and is a federally designated 501(c)3 nonprofit organization. ALBC is supported by membership dues, gifts from individuals, grants, contracts, and sales of publications and merchandise. You are invited to become a member of the American Livestock Breeds Conservancy and a partner in rare breeds conservation. For information, contact ALBC, PO Box 477, Pittsboro, NC 27312 USA, (919) 542-5704. ❖

Selected Bibliography

Alderson, Lawrence. 1994. *A Chance to Survive.* (Yelvertoft, Northamptonshire: Pilkington Press).

Aldington, Araminta. 1989. *A History of the Jacob Sheep.* (Ashford, Kent: Geerings of Ashford, Ltd.).

Baker, Sue. 1993. *Survival of the Fittest: A Natural History of the Exmoor Pony.* (Exeter: Exmoor Books).

Bixby, D. E., C. J. Christman, C. J. Ehrman, D. P. Sponenberg. 1994. *Taking Stock: The North American Livestock Census.* (Blacksburg, Virginia: McDonald and Woodward Publishing Company).

Briggs, Hilton M. 1969. *Modern Breeds of Livestock.* (3rd edition) (New York: The MacMillan Company).

Budiansky, Stephen. 1995. *The Covenant of the Wild.* (2nd edition) (Leesburg, Virginia: The Terrapin Press).

Clutton-Brock, Juliet. 1987. *A Natural History of Domesticated Mammals.* (London: Cambridge University Press and British Museum of Natural History).

Coburn, F. D. 1916. *Swine in America.* (New York: Orange Judd).

Fitzhugh, H. A. and G. E. Bradford, ed. 1983. *Hair Sheep of Western Africa and the Americas.* (Boulder, Colorado: Westview Press).

Fournier, Nola and Jane Fournier. 1995. *In Sheep's Clothing: A Handspinner's Guide to Wool.* (Loveland, Colorado: Interweave Press).

Fowler, Cary and Pat Mooney. 1990. *Shattering: Food, Politics, and the Loss of Genetic Diversity.* (Tucson, Arizona: University of Arizona Press).

Hall, Stephen J. G. and Juliet Clutton Brock. 1989. *200 Years of British Farm Livestock.* (London: British Museum of Natural History).

Hendricks, Bonnie. 1995. *The International Encyclopedia of the Horse.* (Norman, Oklahoma: University of Oklahoma Press).

Kalk, Bruce. 1991. "The Extinction of Historic Breeds of Swine," *ALBC News* 8(3).

Mason, Ian. 1996. *A World Dictionary of Livestock Breeds, Types, and Varieties.* (Wallinford, Oxon: CAB International).

Plumb, Charles S. 1920. *Types and Breeds of Farm Animals.* (Boston: Ginn and Company).

Porter, Valerie. 1993. *Pigs: A Handbook to the Breeds of the World.* (Ithaca, New York: Cornell University Press).

Ryder, Tom. 1979. *The High Stepper.* (London: J. A. Allen).

Shepherd, S. M. 1886. *The Hog in America: Past and Present.* (Indianapolis, Indiana: Swine Breeders Journal).

Sponenberg, D. Phillip and Carolyn J. Christman. 1995. *A Conservation Breeding Handbook*. (Pittsboro, North Carolina: American Livestock Breeds Conservancy).

Telleen, Maurice. 1977. *The Draft Horse Primer*. (Emmaus, Pennsylvania: Rodale Press).

Tyler, Catherine. 1996 "Changing Habitats and the 20th Century Horse," *ALBC News* 13(1).

Vaughan, Henry W. 1945. *Breeds of Livestock in America.* (Columbus, Ohio: Long's College Book Company).

Breed and Species Organizations

Asses

American **Donkey & Mule** Society, 2901 N. Elm, Denton, TX 76201, (940) 382-6845

Canadian **Donkey & Mule** Association, R.R. 10, Brampton, ON L6V 3N2 Canada

American Mammoth Jackstock Registry, 6513 W. Laurel Road, London, KY 40741-9717, (606) 878-0486

International Registry of **American Mammoth Jackstock**, P.O. Box 38, Floyd's Knob, IN 47119, (502) 894-9830

American Mammoth Jackstock, also contact the American Donkey and Mule Society (see above)

North American Baudet du **Poitou** Society, c/o Altamont Farm, P.O. Box 1127, Millbrook, NY 12545, (914) 677-8376

Cattle

Ankole-Watusi International Registry, 22484 W. 239th St., Spring Hill, KS 66038-9306, (913) 592-4050

Ayrshire Breeders Association, P.O. Box 1608, Brattleboro, VT 05302-1608, (802) 254-7460

Belted Galloway Society, 174 Spore Road, Potts Camp, MO 38659, (601) 333-4172

Societe des Eleveurs de Bovins **Canadienne**, 68 Dolbeau, Sherbrooke, Qc J1G 2Z7 Canada, (819) 346-1258

Devon Cattle Association, Inc., c/o Bernard Mayeux, Rt. 1, Box 93, Bunkie, LA 71322, (318) 876-3630

America **Dexter** Cattle Association, Rt. 1 Box 378, Concordia, MO 64020, (816) 463-7704

Canadian **Dexter** Cattle Association, R.R. 1, Keene, ON K0L 2G0 Canada, (705) 295-6536

Dutch Belted Cattle Association of America, c/o the American Livestock Breeds Conservancy, P.O. Box 477, Pittsboro, NC 27312, (919) 542-5704

Florida Cracker Cattle Association, c/o Dr. T.A. Olson, Department of Animal Science, University of Florida, P.O. Box 110910, Gainesville, FL 32611-0910, (904) 392-2367

American **Galloway** Breeders Association, c/o Susan Waples, Box 544, Terry, MT 59349, (406) 637-2114 and c/o Bob Mullendore, 310 W. Spruce, Missoula, MT 59802, (406) 728-5719

Canadian **Galloway** Association, R.R. 1, Puslinch, ON NOB 2JO Canada, (519) 821-7493

American **Guernsey** Cattle Club, P.O. Box 666, Reynoldsburg, OH 43068-0666, (614) 864-2409

Canadian **Guernsey** Association, 368 Woolwich St., Guelph, ON N1H 3W6 Canada, (519) 836-2141

American **Highland** Cattle Association, 4701 Marion St., Suite 200 Exchange Building, Denver, CO 80216, (303) 292-9102

Canadian **Highland** Cattle Society, 307 Spicer, Knowlton, Qc JOE 1VO Canada, (514) 243-5543

North American **Kerry** Cattle Association, c/o

Rare Breeds Canada, Tom Hutchinson, Trent University, Environment/Resources Studies, Box 4800, Peterboro, ON K95 7B8 Canada, (705) 748-1634

American **Milking Devon** Association, P.O. Box 730, New Durham, NH 03855, (603) 859-6611

American **Milking Shorthorn** Society, P.O. Box 449, Beloit, WI 53511-0449, (608) 365-3332

Pineywoods cattle, contact the American Livestock Breeds Conservancy, P.O. Box 477, Pittsboro, NC 27312, (919) 542-5704

Randall Lineback cattle, contact the American Livestock Breeds Conservancy, P.O. Box 477, Pittsboro, NC 27312, (919) 542-5704

American **Red Poll** Association, P.O. Box 014096, Kansas City, MO 64010, (816) 842-3512

Canadian **Red Poll** Association, R.R. 3, Ponoka, AB TOC 2HO Canada, (403) 783-5951

North American **White Park** Cattle Association, HC 87 Box 2214, Big Timber, MT 59011, (406) 932-4197

Goats

American **Goat** Society Inc., R.R. 1 Box 56, Esperance, NY 12066-9704, (518) 875-6708

American **Dairy Goat** Association, P.O. Box 865, Spindale, NC 28160, (704) 286-3801

International **Dairy Goat** Registry, Rt. 1, Box 265-A, Rossville, GA 30741, (706) 375-4326

American **Nigerian Dwarf** Organization, 1433 Glocker Road, Sedgwick, KS 67135, (316) 772-0297

American **Nigerian Dwarf** Dairy Association, 8090 Mercury, Flagstaff, AZ 85027, (520) 526-8956

Nigerian Dwarf Goat Association, 1523 Jewett Lane, DeSoto, TX 75115, (972) 274-1740

Nigerian Dwarf, also contact the American Goat Society and the International Dairy Goat Registry (above)

Oberhasli Breeders Association, 11620 Sunset Ct., Montague, CA 96064, (916) 459-3999; and the American Dairy Goat Association (above)

San Clemente goat, contact the American Livestock Breeds Conservancy, P.O. Box 477, Pittsboro, NC 27312, (919) 542-5704

Spanish goat, contact the American Livestock Breeds Conservancy, P.O. Box 477, Pittsboro, NC 27312, (919) 542-5704

American **Tennessee Fainting** Goat Association, R.R. 4, Box 4100, Houlton, ME 04730

International **Fainting** Goat Association, 3450 230th Street, Terril, IA 51364, (712) 853-6372

Horses

American **Horse** Council, 1700 K Street, NW, Suite 300, Washington, DC 20006, (202) 296-4031

Akhal-Teke Registry of America, Rt. 5, Box 110, Staunton, VA 24401, (540) 886-1870

American Cream Draft Horse Association, 2065 Noble Ave, Charles City, IA 50616-9108, (515) 228-5308

Societe des Eleveurs de Chevaux **Canadians**, CP 38, Succursale C, Montreal, Qc H2L 3J7 Canada, (514) 596-0833

Caspian Horse Society, 4125 Arrowhead NE, Oklahoma City, OK 73078, (888) 327-4871

Cleveland Bay Horse Society of North America, P.O. Box 221, S. Windham, CT 06266, (860) 423-9457

Clydesdale Breeders of the United States, 17378 Kelly Road, Pecatonica, IL 61063, (815) 247-8780

Clydesdale Horse Association of Canada, R.R. 2, Thornton, ON L0L 2N0 Canada, (705) 458-9214

Dartmoor Pony Society of America, 145 Upper Ridgeview Road, Columbus, NC 28722, (704) 894-5672

American **Dartmoor** Pony Association, 15870 Pasco-Montra Road, Anna, OH 45302, (513) 596-6623

Canadian Mountain and Moorland Society (**Exmoor** ponies), c/o Anne Holmes, P.O. Box 155, Ripley, Ontario N0G 2R0 Canada, (519) 395-4273

Florida Cracker Horse Association, Inc., P.O. Box 186, Newberry, FL 32669, (904) 472-2228

Friesian Horse Association of North America, P.O. Box 11217, Lexington, KY 40574-1217, (541) 549-4272

Gotland Horse Registry (USA), c/o the American Livestock Breeds Conservancy, P.O. Box 477, Pittsboro, NC 27312, (919) 542-5704

American **Hackney** Horse Society, 4059 Iron Works Pike, Building A, Lexington, KY 40511, (606) 255-8694

Hackney Horse Affiliates, 8999 Fairmount Road, Novelty, OH 44072, (216) 338-6301

Canadian **Hackney** Society, R.R. 1, Lindsay, ON K9V 4R1Canada, (705) 324-5277

Lipizzan Association of North America, P.O. Box 1133, Anderson, IN 46015, (317) 644-3904

U.S. **Lipizzan** Registry, c/o NELA, 707 13th Street SE, Suite 275, Salem, OR 97301, (503) 589-3172

Mountain Pleasure Horse Association, P.O. Box 670, Paris, KY 40362, (606) 987-1695

Rocky Mountain Horse Association, 1140 McCalls Mill Road, Lexington, KY 40515, (606) 263-4374

American **Shire** Horse Association, 35380 County Road 31, Davis, CA 95616, (916) 757-2742

Spanish Barb Breeders Association, 13862 Gulf Haven Road, Gulfport, MS 39503-5173, (601) 831-5173

American Indian Horse Registry (**Spanish** horses) Rt. 3, Box 64, Lockhart, TX 78644, (512) 398-6642

Southwest **Spanish Mustang** Association, P.O. Box 48, Finley, OK 74543

Spanish Mustang Registry, HCR 3, Box 7670, Willcox, AZ 85643, (602) 384-2886

American **Suffolk** Horse Association, 4240 Goehring Rd, Ledbetter, TX 78946-5004, (409) 249-5795

Pigs

Choctaw pig, contact the American Livestock Breeds Conservancy, P.O. Box 477, Pittsboro, NC 27312, (919) 542-5704

Gloucestershire Old Spots Registry, 31010 Peterson Rd, Philomath, OR 97370, (541) 929-5487

Guinea Hog Registry, 14335 Pauma Vista, Valley Center, CA 92082, (619) 749-2126

National **Hereford** Hog Record Association, Rt. 1, Box 37, Flandreau, SD 57028, (605) 997-2116

Large Black pig, contact the American Livestock Breeds Conservancy, P.O. Box 477, Pittsboro, NC 27312, (919) 542-5704

Mulefoot pig, contact the American Livestock Breeds Conservancy, P.O. Box 477, Pittsboro, NC 27312, (919) 542-5704

Ossabaw Island pig, contact the American Livestock Breeds Conservancy, P.O. Box 477, Pittsboro, NC 27312, (919) 542-5704

Wenglar's **Red Waddle** Hog Association, Rt. 7 Box 153, Jacksonville, TX 75766, (903)586-6138, or contact the American Livestock Breeds Conservancy, P.O. Box 477, Pittsboro, NC 27312, (919) 542-5704

American **Saddleback** Registry, Box 41, Eldridge, AL 35554, (205) 924-8916, or contact the American Livestock Breeds Conservancy, P.O. Box 477, Pittsboro, NC 27312, (919) 542-5704

Tamworth Swine Association, 200 Centenary Road, Winchester, OH 45697, (937) 695-0114

Sheep

American **Sheep** Industry Association, 6911 S. Yosemite St, Englewood, CO 80112, (303) 771-3500

Information on **sheep** breeds in Canada is available from the Canadian Livestock Records Association, 2417 Holly Lane, Ottawa, ON K1V 0M7, (613) 731-7110

Barbados Blackbelly Sheep Association USA, 18412 E. Indian Hill Road, Newalla, OK 74857, (405) 360-5187 or contact the American Livestock Breeds Conservancy, P.O. Box 477, Pittsboro, NC 27312, (919) 542-5704

American **CVM-Romeldale** Registry, P.O. Box 630, Carnation, WA 98014, (206) 333-4934

American **Cotswold** Record Association, P.O. Box 59, Plympton, MA 02367, (617) 585-2026

Black Cotswold Society, P.O. Box 542, Goffstown, NH 03045, (603) 774-4104

Cotswold Breeders Association, 15908 E. Willis Rd, Gilbert, AZ 85296, (620) 988-3832

Continental **Dorset** Club, P.O. Box 506, Hudson, IA 50643, (319) 988-4122

Gulf Coast Native Sheep Registry, c/o the American Livestock Breeds Conservancy, P.O. Box 477, Pittsboro, NC 27312, (919) 542-5704

Gulf Coast Sheep Breeders Association, 2022 Cedar Canyon, Ft. Wayne, IN 46845, (219) 637-6369

Hog Island Sheep Registry, c/o the American Livestock Breeds Conservancy, P.O. Box 477, Pittsboro, NC 27312, (919) 542-5704

Jacob Sheep Breeders Association, 6350 E. County Road 56, Fort Collins, CO 80524-9340, (970) 484-3344

Jacob Sheep Conservancy, 1165 E. Lucas Road, Lucas, TX 75002, (972) 727-0900

American **Karakul** Sheep Registry, 3026 Thomas Road, Rice, WA 99167, (509) 738-6310

Katahdin Hair Sheep International, P.O. Box 115, Fairview, KS 66425, (913) 467-8041

Leicester Longwool Breeders Association, c/o Colonial Williamsburg Foundation, Coach and Livestock Department, P.O. Box 1776, Williamsburg, VA 23187, (757) 220-7790

Navajo-Churro Sheep Association, Taos Junction, Box 94, Ojo Caliente, NM 87549, (505) 707-0088

American **Oxford** Sheep Association, 1960 E. 2100 N. Road, Stonington, IL 62567, (217) 325-3515

Santa Cruz Island Sheep Registry, c/o the American Livestock Breeds Conservancy, P.O. Box 477, Pittsboro, NC 27312, (919) 542-5704

North American **Shetland** Sheep Breeders Association, 265 Truway Rd., Luxemburg, WI 54217, (902) 837-2167

St. Croix Sheep Breeders Association, c/o Dr. Cole Evans, Utah State University, UMC 4815, Logan, UT 84322-4815, (801) 797-2181

National **Tunis** Sheep Registry, 311 Scotch Settlement Rd., Gouverneur, NY 13642, (315) 287-3776

Wiltshire Horn Sheep Registry, c/o the American Livestock Breeds Conservancy, P.O. Box 477, Pittsboro, NC 27312, (919) 542-5704

Photo Credits

Cover

Dexter cow - Catherine Palladino
Hereford pig - Mark Hess
American Mammoth Jackstock - Mary Ellen Nicholas
San Clemente goat - Richard Kirschman
Navajo-Churro sheep - Maryanne Mott
Gotland horse - Karen Kachur

Chapter 1. Why Rare Breeds?

5 American Mammoth Jackstock - Mary Ellen Nicholas

Chapter 2. Asses

7 Poitou - Deborah Hamilton
9 American Mammoth Jackstock - Mary Ellen Nicholas
11 Poitou - Deborah Hamilton

Chapter 3. Cattle

13 Dutch Belted - Catherine Paladino
15 Ankole-Watusi - courtesy of Ankole-Watusi International Registry
16 Ayrshire - Judy Black
18 Belted Galloway - Gene Hebert
19 Canadienne - courtesy of Canadienne Cattle Breeders Association
20 Devon - ALBC
21 Dexter - Maryanne Mott
22 Dexter and Milking Shorthorn - Maryanne Mott
23 Dutch Belted - Kenneth Hoffman
24 Florida Cracker - Phillip Sponenberg
26 Galloway - Patricia Pruitt
27 Guernsey - Elizabeth Henson
29 Highland - Richard Nicoll
30 Kerry - Don Bixby
31 Milking Devon - Rob Amberg
32 Milking Shorthorn - Don Bixby
34 Pineywoods - Phillip Sponenberg
35 Randall Lineback - Cynthia Creech
36 Red Poll - Elizabeth Henson
38 White Park - Don Bixby

Chapter 4. Goats

39 San Clemente - Sheri Tepper
41 Nigerian Dwarf - Mary Ellen Nicholas
42 Oberhasli - courtesy of Peggy and Dave Daubert
43 San Clemente - Richard Kirschman
44 Spanish - Phillip Sponenberg
46 Tennessee Fainting - Phillip Sponenberg

Chapter 5. Horses

47 Suffolk - Maryanne Mott
48 Akhal-Teke - Cathy Leddy
50 American Cream Draft - Colonial Williamsburg Foundation
51 Canadian - courtesy of Ginny Dailly-Sealy
52 Caspian - courtesy of Patrick Carmack
53 Cleveland Bay - David Field
55 Clydesdale - courtesy of Clydesdale Breeders of the United States
56 Dartmoor - Cynthia Brann
57 Exmoor - Linda Lucas
58 Florida Cracker - Phillip Sponenberg
59 Friesian - Catherine Tyler
60 Gotland - Karen Kachur
62 Hackney - Phillip Sponenberg
63 Lipizzan - Catherine Tyler
64 Mountain Pleasure - courtesy of the Mountain Pleasure Horse Association
65 Rocky Mountain - courtesy of the Rocky Mountain Horse Association
66 Shire - Sharon McLin
68 Spanish Mustang/Spanish Barb - Marye Ann Thompson
69 Suffolk - Maryanne Mott

Chaper 6. Pigs

71 Red Wattle/Tamworth hogs - Don Bixby
73 Choctaw - Jim Combs
74 Gloucestershire Old Spots - Don Bixby
76 Guinea Hog - Marcia Read
77 Hereford - Mark Hess
78 Large Black - C. S. Lomas

79 Mulefoot - J. Gordon Miller
80 Ossabaw Island - Don Bixby
81 Red Wattle - M. Prentice
82 Saddleback - Elizabeth Henson
83 Tamworth - Don Bixby

Chapter 7. Sheep

85 Leicester Longwool - Colonial Williamsburg Foundation
87 Barbados Blackbelly - courtesy of Nancy Schilb
88 CVM - Lewis White
89 Cotswold - American Cotswold Record Association
90 Dorset Horn - Joe Greene
92 Gulf Coast Native - Don Bixby
93 Hog Island - courtesy of Gunston Hall
94 Jacob (American) - Maryanne Mott
95 Karakul (American) - Jeffrey Black
96 Katahdin - courtesy of Katahdin Hair Sheep International
97 Leicester Longwool - Phillip Sponenberg
99 Navajo-Churro - Tanya Charter, courtesy of Navajo-Churro Sheep Association
100 Oxford - courtesy of Tweed Croft Oxfords
101 Santa Cruz- Joe McCummins
102 Shetland - Linda Zuppann
103 St. Croix - Agricultural Research Service-USDA
104 Tunis - Robert Gear
105 Wiltshire Horn - courtesy of Carol Postley

Chapter 8. Conservation Breeding

107 Man holding Tamworth piglets - Gene Hebert

Index

Akhal-Teke horses, 48-49
American Cream horses, 49-50
American Livestock Breeds Conservancy, 1, 23, 25, 61, 75, 92, 108
American Mammoth Jackstock, 7-9
Ankole-Watusi cattle, 15-16
Asses, 7-11
Ayrshire cattle, 13, 14, 17

Barbados Blackbelly sheep, 87-88
Belted Galloway cattle, 17-18, 26
Breed(s), definition of, 5
  types of, 3
Brush goats, *see* Spanish goats

CVM sheep, *see* California Variegated Mutant/Romeldale sheep
California Variegated Mutant/Romeldale sheep, 88-89
Canadian horses, 47, 51-52
Canadienne cattle, 13, 19-20,
Caspian horses, 52-53
Cattle, 13-38
Choctaw pigs, 72, 73-74
Cleveland Bay horses, 53-54
Clydesdale horses, 54-55
Conservation, reasons for, 1, 6, 14
  how to become involved in, 107-108
  priority breeds list, 2, 4
Cotswold sheep, 89-90

Dartmoor ponies, 56-57
Devon cattle, 20-21, 31-32
  see also Milking Devon cattle
Dexter cattle, 21-22
Donkeys, *see* Asses
Dorset Horn sheep, 90-91
Dutch Belted cattle, 13, 22-24

Exmoor ponies, 57-58

Fainting goats, *see* Tennessee Fainting goats
Feral breed, definition of, 3
Florida Cracker cattle, 13, 24-25
Florida Cracker horses, 58-59
Friesian horses, 59-60

Galloway cattle, 17, 25-26
Gloucestershire Old Spots pigs, 75
Goats, 39-46
Gotland horses, 61
Guernsey cattle, 14, 27-28
Guinea Hogs, 72, 75-77
Gulf Coast Native sheep, 91-92

Hackney horses, 61-62
Hair sheep, 86, 87-88, 96-97, 103-104, 105-106
Hereford pigs, 77-78
Highland cattle, 28-29
Hog Island sheep, 93
Hogs, *see* Pigs
Horses, 47-70

Industrialized agriculture, 2, 3, 6, 72-73
Industrialized stocks, definition of, 3

Jacob sheep (American), 94-95

Karakul sheep (American), 1, 95-96
Katahdin sheep, 96-97
Kerry cattle, 30-31

Landrace, definition of, 3
Large Black pigs, 78-79
Leicester Longwool sheep, 97-98
Lineback cattle, *see* Randall Lineback cattle
Lipizzan horses, 63-64

Milking Devon cattle, 21, 31-32
  see also Devon cattle
Milking Shorthorn cattle, 32-33
Mountain Pleasure horses, 64-65
Mulefoot pigs, 72, 79-80
Myotonic goats, *see* Tennessee Fainting goats

Navajo-Churro sheep, 98-99
Nigerian Dwarf goats, 40-41

Oberhasli goats, 42
Ossabaw Island pigs, 80-81
Oxford sheep, 100

Pigs, 71-84
Pineywoods cattle, 13, 33-35
Poitou asses, 8, 10-11

Randall Lineback cattle, 35-36
Rare breeds, definition of, 2, 4, 5-6
Red Poll cattle, 36-37
Red Wattle pigs, 81-82
Rocky Mountain horses, 65-66
Romeldale sheep, *see* California Variegated Mutant/Romeldale

Saddleback pigs, 82-83
San Clemente goats, 43-44
Santa Cruz sheep, 101-102
Sheep, 1, 85-106
Shetland sheep, 1, 102-103
Shire horses, 67
Spanish goats, 40, 44-45
Spanish Mustang/Spanish Barb horses, 47, 67-69
St. Croix sheep, 1, 103-104
Standardized breed, definition of, 3
Suffolk horses, 69-70
Swine, *see* Pigs

Tamworth pigs, 72, 83-84
Tennessee Fainting goats, 45-46
Tunis sheep, 104-105

White Park cattle, 37-38
Wiltshire Horn sheep, 105-106
Wooden Leg goats, *see* Tennessee Fainting goats
Wool, 85-86